Neurology
Equations
made simple

Neurology Equations made simple

Differential diagnosis and Neuroemergencies

Dr Nadeem Akhtar

authorHOUSE®

AuthorHouse™ UK
1663 Liberty Drive
Bloomington, IN 47403 USA
www.authorhouse.co.uk
Phone: 0800 047 8203 (Domestic TFN)
*   +44 1908 723714 (International)*

Published by AuthorHouse 02/20/2020

ISBN: 978-1-5049-9027-1 (sc)
ISBN: 978-1-5049-9028-8 (hc)
ISBN: 978-1-5049-9029-5 (e)

To my father Sehba Akhtar, my mother Saeeda Akhtar, my sister Rubina Akhtar and rest of my family for their *endless support which has left fingerprints on my heart.*

Contents

Movement Disorders

Ataxia of sub acute onset

Post Infectious Cerebellitis	▪ Acute or subacute onset, following an infection usually viral; (can be GI or, pulmonary).
	▪ Ataxia can develop over several hours, days or, 1-4 weeks after infection.
	▪ EBV, influenza, parainfluenza, HSV, VZV, CMV, adenovirus, HIV, mumps - Lyme (B. burgdorferi) – Mycoplasma – Legionella can be involved
	↪ Neurological Equation Sudden onset of unsteadiness (truncal ataxia + Hx of diarrhoea/ chest infection + intact DTR (If lost think of MFS) = Post infectious Cerebellitis
Vascular Cerebellar Haemorrhage/ Infarction	▪ Long-standing hypertension ---most common cause
	▪ Anticoagulant use
	▪ Heamorrahge into tumor
	↪ Neurological Equation Sudden abrupt onset headache + Nausea and vomiting + Inability to walk (truncal ataxia) + Loss or alteration of consciousness = Cerebellar haemorrhage, (infarction will be without headache)
Miller Fisher Syndrome	▪ The Miller Fischer syndrome which is an acute self limiting condition, is characterised by truncal ataxia, areflexia and opthamopleigia
	▪ The anti–Gqib Igg antibody titier is most commonly elevated in MFS, Gullain Barre syndrome and Bickerstaff encpaejalitis.

	• The ataxia is similar to cerebellar disease, but it is not yet known whether it arises centrally or peripherally. 🖑 Neurological Equation Sudden onset of unsteadiness (truncal ataxia + Diplopia (complex opthalmoplegia) + loss of DTR + anti-GQ1b IgG antibody = MFS)
Multiple Sclerosis	• Multiple sclerosis can produce disorders of equilibrium of cerebellar, vestibular, or sensory origin. • More common in women • Gait and limb ataxia (gait involvement is more common than limb.
Paraneoplastic Subacute cerebellar degeneration	• Severe dysarthria • Common tumors which cause Paraneoplastic sub-acute cerebellar degeneration are • lung carcinoma (small-cell) • ovarian uterine • lymphomas 🖑 Neurological Equation Known case of cancer (can precede the identification of the cancer) + Subacute onset of Ataxia/dysarthria + Anti Hu antibodies/Anti Yo = Paraneoplastic Subacute cerebellar degeneration
Pearls	🖑 **Other known Paraneoplastic Syndromes** • Lambert-Eaton myasthenic syndrome (LEMS) • Opsoclonus/myoclonus • Sensory neuronopathy

Dr Nadeem Akhtar

Creutzfeldt-Jakob disease	▪ In CJD, cerebellar signs are present in about 60% and in some patients present with isolated ataxia in about 10% of cases.
	▪ Although cognitive compromise is more prominent clinical feature.
Other causes	▪ Infectious Cerebellar abscess.
	▪ Viral encephalitis
	▪ Cerebral vasculitis

Chronic Ataxia

Friedreich ataxia	<ul><li>Friedreich ataxia is a slowly progressive ataxia with mean age of onset between age 10 and 15 years and commonly before age of 25 years.</li><li>25% can have a typical age of presentation which could be after 25 years of age.</li><li>Associated features include weakness in pyramidal distribution or extensors planters.</li><li>Peripheral neuropathy (axonal)</li><li>Optic atrophyhy (25%)</li><li>Skeletal deformities: pes cavus (50%) and scoliosis (75 %)</li><li>Hypertrophic Cardiomyopathy 25%</li><li>Diabetes or glucose intolerance (20%)</li></ul> ↳ Neurological Equation Pt under 25 years of age + Cerebellar signs + Pyramidal signs (up going planters) + peripheral neuropathy (absent ankle jerks) + Optic atrophy = Friedreich ataxia
Pearls	↳ **D/D of extensor plantars with absent ankle jerks includes**<ul><li>FA ,</li><li>MND</li><li>Tabes Dorlasis</li></ul>

Dr Nadeem Akhtar

SCA	There are more than 20 types of SCA These types are given numbers (1 to 22, excluding the number 9). Spinocerebellar ataxias (SCAs) is characterised by progressive cerebllar ataxia in association with - Ophthalmoplegia - Pyramidal signs - Pigmentry retinopathy - Peripheral neuropathy - dementia
Ataxia with Isolated vitamin E deficienciy	Examples of treatable ataxias include those due to deficiencies of vitamin E or coenzyme Q10 and Episodic Ataxia Type 2 (EA2) - Clinical phenotype similar to Friedreich ataxia - Onset is between 2-52 years and generally < than 20 yrs. - Slowly progressive. - Skin signs can be xanthelasmata and tendon xanthomas. 🖎 Neurological Equation Ataxia + hint of malabsorption (hx of diarrhea/anaemia/low calcium) + skin xantlemas = Ataxia with isolated vitamin –E deficiency 🖎 **Causes of deficiency include** - Cystic fibrosis, - Abetalipoprtenemia - Short bowel syndrome - Malabsorption syndrome

Alcohol	▪ Degenerative changes in the cerebellum are largely restricted to the superior vermis ▪ probably as a result of nutritional deficiency. ▪ Affected patients typically have a history of daily or binge drinking with associated dietary inadequacy ↬ Neurological Equation HX of Alcohol overuse + progressive gait disturbance (predominant involvement of vermis) + minimal limb ataxia (relative sparing of cerebellar hemispheres) + MRI showing topical vermial atrophy = Alcohol related cerebellar degeneration
Ataxia telangiectasia	▪ Slowly progressive ataxia and dysarthria associated Choreoathetosis ▪ Look for, cutaneous, conjuctival and bulbar telangiectasia ▪ Bulbar conjunctivae are typically affected first, followed by sun-exposed areas of the skin, including the ears, nose, face, and antecubital and popliteal fossae ▪ Increased susceptibility to infections due (selective Ig A and IgG deficiency) ↬ Neurological Equation Ataxia + recurrent infections (respiratory due to immuno deficiency) + Telangiectasia (one has to look for in conjucntival and bulbar areas) + progressive choreoathetosis +Ataxia(uncommon)=Telangiectasia Ataxia

Adrenloeukodystrophy	▪ Adrenoleukodystrophy is an X-liked hereditary disorder caused by genetic problem in Xq28 affecting the peroxismal membrane. The genetic defect would result in peroxisomal dysfunction resulting in very long –chain fatty acids (VLCFA)
	▪ The nervous system, adrenal cortex, and testis are involved
	▪ Acanthocytes on peripheral blood smears
	▪ Decreased serum cholesterol elvel
	▪ Increased high density lipoprotein cholesterol levels
	↳ Neurological Equation Boy / man (x linked disorder) + Ataxia + cognitive compromise + progressive loss of vision + hypotension and hyper pigmentation (due to adrenal insufficiency)+ increased VLCFA = Adrenoleukodystrophy
Pearls	↳ **Chronic Ataxia with pyramidal signs** ◆ FA ◆ SCA
	↳ **Chronic Ataxia with Extra pyramidal signs** ◆ MSA-C ◆ PSP
	↳ **Ataxia With Progressive Myoclonic Epilepsy** ◆ Unverricht-Lundborg Disease ◆ Lafora Body DSisease ◆ Myoclonic Epilepsy with ragged red fibres

Acute Chorea

Sydenham chorea	<ul><li>It is one of the most common cause of acute chorea in paediatric age group</li><li>Both motor and nonmotor features. are prominent</li><li>tics can accompany chorea</li><li>Non-motor features include obsessions, compulsions, and emotional lability,</li><li>It is a self limiting disorder with remission within 6-9 months. Recurrence can occur 20-50% of patients.</li><li>investigation: ASO titres, throat culture, cardiac examination</li></ul> ☝ Neurological Equation Paediatric age group + Hx of throat infection (may not be obvious as can be up to few months ago) + emotional labiality (crying or laughing inappropriately + chorea = Sydenham chorea
PANDAS (Paediatric autoimmune neuropsychiatric disorders associated with streptococcal throat infections)	<ul><li>Both Sydenham chorea and PANDAs haven neuropsychiatric symptoms and differentiate one form the other at the onset cane be challenging.</li></ul>
Neuroacanthocytosis	<ul><li>A slow neurodegenerative condition which involves predominantly of the basal ganglia, and erythrocyte acanthocytosis.</li></ul>

Dr Nadeem Akhtar

	▪ The clinical presentation typically has both chorea and dystonia, but other movement disorders may be seen
	↳ Neurological Equation Patient in 30s + Chorea or any form of Movement disorder (dystonia. Myclonus, tics)+/- PD like features) + Ataxia + orolingual dystonia + acanthocytes (> 3 % is abnormal) = Neuroacanthocytosis
Pearls	▪ To check for acnthoycytes one has to send fresh blood sample at least three time. Warn the lab what you are looking for.
Anti-phospholipid syndrome Systemic Lupus Erythematosus Vascular causes	▪ Stroke- infarction of subthamaic nucleus can result in hemichorea
Chorea gravidarum.	▪ Sudden onset of chorea generally during first trimester pregnancy,
	▪ the incidence is 1 by each 2000pregnancies,
	▪ it is self limiting condition which resolves with the pregnancy.
	▪ The recurrences may occur in the future pregnancies
Pearl	▪ **There are few case reports of women developing chorea while being taking oral contraceptive pills**
	↳ Neurological Equation Female +Intermittent Involuntary semi purposeful movements+ Known pregnancy (if not do ask or even organise a pregnancy test if in doubt) = Chorea gravidarum.

Drug-induced chorea
Rare
Post-cardiac surgery
Encephalitis
HIV infection
carbon monoxide

Chronic Chorea

Huntington disease (HD)	• Huntignton disease is a rare condition characterised by
	• chorea, behavioural and Neuro-psychiatric features and progressive dementia
	• Age of onset: 35 - 45 yr
	• Juvenile variant – 5-10% of cases
	• Survival is over a period of 10 – 20 years
	✏ Neurological Equation Paediatric age group + Hx of throat Chorea+ dystonia and parkinsonian features + Dementia and the psych-iatric irritability, unsteadiness= HD
Wilson disease	• Signal abnormalities are bilateral and involve basal ganglia , thalamus and parietal lobe
	• Low ceruloplasmin and low serum copper
	• Elevated 24hour urine copper
	• Kayser-Fleisher rings.(visible on slit lamp examination and can be missed by naked eye)
	✏ Neurological Equation Chroea + Extrapyrmidal features + Deranged LFTs (in adavcned cases cirrhosis of liver) = Wilson disease
Benign hereditary chorea	• Onset characteristically is in early childhood (unlike HD) ,
	• Non-progressive with normal life span
	• No dementia,

Dentatorubralpalli doluysian atrophy (DRPLA	▪ DRPLA-Dentatorubral-pallidoluysain atrophy is a rare autosomal dominant, neurodegerative disorder Clinical features include ▪ epilepsy, ataxia myoclonus, choreiform ▪ movements, and dementia, can include psychiatric disease ✍ Neurological Equation chorea + dementia (these features are similar to HD) BUT + ataxia +seizure + Myoclonus (should alert about this condition) = DRPLA
Senile chorea Familial chorea-ataxia-myoclonus syndrome Polycythemia vera Spinocerebellar ataxias Paroxysmal kinosegenic choreoathetosis	

Acute Dystonia

Narcoleptic therapy	<ul><li>Neuroleptics having dopamine receptor-blocking properties are frequently responsible for the development of movement disorders.</li><li>Parkinsonism, acute dystonic reactions, akathisia, and tardive dyskinesias are the various clinical presentations</li><li>Generally it becomes evident after few days of imitation (90% within 5 days of therapy),</li><li>Dystonic reactions, can range from short and intermittent brief jerks to prolonged muscle spam involving the carnio-cervical region.</li></ul> ✤ Neurological Equation Use of neuroleptics (just few days after intiation) + abnormal posturing = acute dystonic reaction
Neuroleptic-malignant syndrome	
Sub acute or chronic Dystonia	
Dystonia Overview	<ul><li>Dystonia is cahraterised by suatianed muscle contraction of one or more sites of the body are involved.</li><li>Dystopia can also cause unusual snaking and repetitive movements and or abnormal postures</li><li>Dystonia sometimes is associated with dystonic tremor</li><li>Some dystonic movements are aggravated by movement, called "action dystonia". They may be nonspecific or task-specific (e.g., writing,)</li></ul>

Primary dystonia	Dystonia + • no structural abnormality in the CNS + no nuerodegenration = Primary dystonia (often genetic) ■ Characteristically it has a childhood onset ■ Initially focal , mainly foot; ■ This can progress to become segmental or even generalised. ■ In more severe cases, some of the body parts may remain in sustained fixed postures ■ Cranio-cervical muscles are usually spared ■ Primary dystonia are not associated with any laboratory abnormalities ↬ Neurological Equation Dystonia+ focal in onset+ in a young patient(28 year of age) + secondary generalization (slowly) + (but) typical sparing of neck= Primary dystonia (> 60 have DYT1 mutation)
PEARL	■ Imaging and investigations will be normal Not associated with any neurodegenration
Dopa-responsive dystonia	↬ Neurological Equation Primary dystonia (a sub type of primary dystonia with no structural abnormality in the CNS + no nuerodegenration) + childhood onset (generally age) therapeutic response to L dopa = Dopa-responsive dystonia
	Typical history and pattern of onset with a positive family ■ Another subtype of primary dystonia as with other types of primary dystonia the primary movment disorder is dystonia with no other prominent signs)

Dr Nadeem Akhtar

Adult-onset primary dystonias	▪ Aetiology of primary adult –onset focal dystonia is usually remains unknown.
	▪ This condition generally occurs sporadically in the absence of family history similar disorder.
	▪ Feet are affected uncommonly in adult onset primary dystonia unlike childhood-onset dystonia in which feet are typically involved at the onset .
	↳ Neurological Equation Priamry Dystonia + adult onset (over 25 years of age unlike childhood which is usually)+ focal onset (but sparing of feet is typical , unlike child hood onset which involves feet) = Adult-onset primary dystonias (cause is generally not genetic but extensive test should be done to rule out secondary causes)
Pearl	▪ If dystonia affects the foot in an adults then one should look for a secondary cause.
	▪ This disorder typically presents in childhood with dystonia of the limb and commonly it is associated with Parkinsonism and occasionally spasticity.
Dystonia plus syndromes	▪ Dystonia-plus syndromes are characterised by dystonia is accompanied by other neurological features.
	▪ Diurnal fluctuation is also a characteristic feature and there will be gradual worsening of symptoms generally occur throughout the day. However this typical feature is present in only 60 % of cases

	✏ Neurological Equation Dystonia (diurnal fluctuation)+ Pyramidal signs +/_ Cerebellar signs +/-Peripheral neuropathy +/- eye signs (optic atrophy, retinitis pigmentosa) = Dystonia plus syndromes
Heredodegenerative Dystonia	✏ Neurological Equation Dystonia +underlying brain degeneration = Heredodegenerative Dystonia - Parkinson's Plus Syndromes - Mitochondrial (Leigh's disease, Leber's) - Wilson's Disease - Huntington's (Westphal variant) - Spinocerebellar Ataxia - Neuroacanthocytosis
Secondary Dystonia	✏ Neurological Equation Dystonia +demonstrable exogenous or structural (trauma, infectious process, birth injury, or developmental delay = Secondary Dystonia
Hemidystonia	Dystonia secondary to structural damage of basal ganglia in particular putamen, that is due
Psychogenic dystonia	- Tumour - Haemorrhage - stroke
Pearl	- Adult-onset Early cranial or bulbar involvement before the lower limbs

Dr Nadeem Akhtar

Myoclonus

With epilepsy syndromes

Juvenile myoclonic epilepsy

Juvenile myclonus epilepsy (JME) is a common epileptic syndrome,

- Age of onset is 6 through 22 years of age,

- Myoclonic jerks are often associated with generalized tonci- clonic seizures and sometime with absence seizures.

- Characteristic myoclonic seizures and GCTC seizures occur in about 90 % of patients, and absence seizures in about 30 % .

- Interestingly, there are no intellectual impairment and there is no abnormalities on clinical examination.

- EEG shows generalized poly-spike and waves which are quite typical pattern in this condition.

- ✌ Neurological Equation
  mycolonic jerks + gtcs + juvenile age group + (characteristic absence of other neurological signs ie ataxia, dementia ect, if present it is PME not JME)

Progressive myoclonic epilepsy	▪ Progressive Myoclonic epilepsy is asyndrome which is characterised by features of myoclonus, cognitive compromise, ataxia and some other slowly progressive neurological deficts. Conditions included are
	▪ Lafora body Disease
	▪ Myoclonic epilepsy with ragged-red fibre (MERRF) syndrome
	▪ Progressive Myoclonic epilepsy is a syndrome which is characterised by features of myoclonus , cognitive compromise ,ataxia and some other slowly progressive neurological deficits. Conditions included are
	▪ Lafora body Disease
	▪ Myoclonic epilepsy with ragged −red fibre (MERRF) syndrome
	▪ Dentato-rubro-palliidal atrophy
	▪ Unverricht-Lunborg disease
	▪ Neuronal ceroid lipofuscinoses
	✍ Neurological Equation Young age + myoclonic jerks + gradual cognitive decline + (characteristic absence of generalised seizures or absence seizures- which is feature of JME) + ataxia = PME
Myoclonus with dementia	▪ Myoclonus occurs in about 50% of cases of corticobasal degeneration.
	▪ It is usually focal, affecting one arm or less commonly a leg.
	▪ Action and reflex myoclonus, can be induced by sensory stimulation of the affected limb,

Corticobasal ganglionic degeneration	• Should be considered if there is a history of rapidly progressive cognitive decline and myoclonus
	• Myoclonus in appearance of periodic synchorus discharges (PSDs) on a routine electroencephalograms (EEGs
Creutzfeldt-Jakob	• Consider if there is a combination of Myoclonus and progressive ataxia suggests the Ramsay-Hunt syndrome causes are: coeliac disease, mitochondriopathies.
Ramsay-Hunt syndrome (progressive myoclonic ataxia, PMA),	

Tics

Overview	<ul><li>Tics are involuntary movement disorders characterised by sudden onset repetitive stereotyped irresistible body movements mainly involving skeletal and pharyngo-laryngeal muscles (responsible for production of noises and sounds</li></ul>
Gilles de la Tourette syndrome	<ul><li>Onset is usually before age 20 years old</li><li>Involves Phonic" or "Vocal"</li><li>Simple phonic tics</li><li>Single, meaningless sound or noise</li><li>Complex phonic tics meaningful utterances and verbalizations</li></ul>
Transient Tic Disorder	<ul><li>Phonic" or "Vocal" have both or only one tic form Duration: 4 weeks to 12 months</li></ul>
Huntington disease Neuroacanthocytosis Autism Schizophrenia	

Dr Nadeem Akhtar

Differential diagnosis of Parkinson Disease

MSA-A (Previously called Shy-Drager syndrome)	- Is a degenerative disorder characterized by parkinsonian features, autonomic insufficiency is also known as multiple system atrophy, - Age of onset is 4th to 6th decades - Symptoms include dizziness, or syncope on standing up, - dysautonomic symptoms. include postural hypotension, anhidrosis, disturbance of sphincter control, impotence. - The prognosis of Shy-Drager syndrome is poor; patients are markedly disabled and have shorter survival. Duration of the illness is 6-7 years - Antiparkinsonian drugs are typically ineffective - ♘ Neurological Equation Extrpyrmidal features (Bradykinesia, rigidty)+ autonomic features (postural drop of > 20 mm hg, incontincnece etc) = MSA
Progressive supranuclear palsy	- Typical for this condition is Early postural instability and tendency to fall backwards - Unlike PD onset of symptoms and signs is symmetrical - resting tremor are rare - blink rate is Markedly decreased (3-5x minute) - Vertical gaze palsy - Astonished facial expression(unlike mask facies of PD)

	▪ Little response to Parkinson's medication ↬ Neurological Equation Symmetrical onset of Extrapyrmidal features (Bradykinesia, rigidity (unlike PD- asymmetrical onset)+ recurrent falls early into disease + supranuclear gaze palsy (restricted vertical gaze in downward dirction) = Progressive supranuclear palsy
Cortical basal ganglionic degeneration	asymmetric cortical and basal ganglionic features which have ▪ Insidious onset ▪ tremor, dystonia, myoclonus
	▪ slow horizontal saccades ▪ alien limb phenomenon ▪ speech impairment; ▪ gait disorder with postural instability; ▪ Cortical dysfunction includes dementia, apraxia, (def)
Creutzfeldt-Jakob disease	▪ May be accompanied by parkinsonian features, but dementia is usually present, ▪ Myoclonic jerking is common, ▪ ataxia is sometimes prominent; and the EEG findings of periodic discharges are usually characteristic.
Normal-pressure hydrocephalus	▪ Traid of gait disturbance (Marche petit pas often mistakenly attributed to parkinsonism), urinary incontinence, and dementia. ▪ CT scanning reveals dilation of the ventricular system of the brain without cortical atrophy

Types of Tremors

	▪ Repetitive, regular oscillatory moves; irregular contraction of opposing muscles,
Rest tremor	▪ Tremors occurring in a bodypart that is completely relaxed and effectively supported against gravity
	▪ Most common cause of resting tremors is parknisnosim but can also occur in advance essential tremors which otherwise mainly an action tremors
	◆ Parkinson diseases ◆ Drug induced ◆ MSA ◆ PSP
Action tremors/Kinetic tremor	▪ Action tremors occur during voluntary contraction of skeletal muscles.
	▪ Kinetic tremors: tremors which occur during guided voluntary movements :writing, touching finger to nose movments
	◆ Physiologic, ◆ Essential tremor ◆ Drug induced ◆ Post-traumatic, ◆ Psychogenic ◆ Cerebellar lesions ◆ Dystonic

Postural tremor	▪ Postural: tremors occurring in a body part maintained against gravity on sustained arm extension. ▪ Eg, tremors occurring in patient holding a newspaper up to read ▪ essential tremors ▪ drug induced, ▪ alcohol withdrawal, ▪ posttraumatic, ▪ psychogenic Physiological trmeors
Essential tremor	▪ The most common type of tremors is essential tremors. ▪ It is an action tremors which is mainly postural (tremors occurring when muscle are active against gravity in a maintained posture) however in some advanced cases it can be kinetic as well as even sporadic rest tremors. To demonstrate ask patients to hold their arms in front against gravity. ▪ ET can also some affect the head , legs and voice. ▪ Interestingly patients with ET do not have any other neurological deficit
Intentional tremor	▪ Intentional tremors are low frequency tremors of about 3-5 hz which are generally disabling for patient ▪ Causes include ▪ MS plaque involving cerebellum ▪ Posterior circulation stroke Tremors in visually guided actions+ increasing in ampliytude upon arrpacing the intended target+ presence of other cerbeallar signs (dysrthrtia, nystagmus ataxia)= intention tremors

Dr Nadeem Akhtar

Orthostatic Tremor	▪ An unusual feeling of and unsteadiness while standing,
	▪ Typically the frequency is of tremors in legs is 13- 18 Hz.
	▪ Characteristic clues in history are that tremor attenuates significantly when patient start to walk or lies down.
	▪ Clinical examination shows a fine tremors which are visible or palpable in the clave muscles
	▪ No helpful treatment is available.
Dystonic movements	▪ Patients with dystonia , the dystonic movements and postures are sometimes accompanied by tremors called dystonic tremors.
	▪ Dystonic tremors typically occur in patients who are younger than 50 years
	▪ It is characteristically irregular and jerky.
	▪ Certain hand pastures can attenuates the tremors
	Tremors are directional + (not oscillating) + jerky and irregular + associated feeling of pulling and discomfort
Psychogenic Tremor	▪ Clinically, distinguishing organic tremors form psychogenic tremors can be challenging .
	▪ Clues suggestive of psychogenic tremors are relatively abrupt onset , unusual fluctuations , extinction and distraction

Facial Movement disorders

Hemifacial spasm	• Typically unilateral, transient but can be recurrent involuntary tonic or clonic movements of muscles facial innervated by facial nerve. • It can be due to abnormal vascular loop compressing the exit zone of facial nerve. ↬ Neurological Equation irregular intermittent spasm of eyelid + spasm of angle of mouth = Hemi facial spasm
Focal seizures	• Partial seizure activity can be localised to face ; in some critically ill patients the activity could be continuous which is called epilepsia partialis continua ↬ Neurological Equation Patients are generally sick (or in ICU) + continuous flickering and contraction of one group of muscles = Focal facial seizures
Facial myokymia	• worm-like contractions or ripple –like quivering ripple-like quivering, mostly involves small areas of face especially around the mouth or eyes, • Associated with multiple sclerosis
Synkinesia	• In patients with Bell's palsy who had aberrant regeneration of facial nerve can have sykinesia which is characterized by subtle, continuous facial movements. • Hx of faicial palsy (Bells palsy) + abnormal subtle movements in muscles of the affected site = Synkinesia

Tics/Facial tics	▪ Tics are habit spasms such as blinking, grimacing, shaking of the head, clearing the throat, coughing or shrugging shoulders,
	▪ worsened by emotion or tiredness.
	▪ abnormal movements of tongue, facial or jaw muscles
	▪ associated with extrapyramidal disease such as athenosis or due to side effects of drugs such as phenothiazines
Facial dyskinesias	↳ Neurological Equation facial + orolingual abnormal movments + hx of use of neurolpetics = Facial dyskinesias

Neuropathies - causes

Acute Axonal Neuropathy

Acute axonal neuropathy

AMAN (Acute motor axonal neuropathy)	• It presents as an acute, flaccid, symmetrical ascending paralysis Weakness reaches peak severity within 5 to 9 days .
	• Bulbar muscles can be involved leading to dysphagia, dysarthria, paralyisis may progress and can lead to quadriplegia and respiratory failure.
	• Autonomic dysfunction is not uncommon
	• Extra-ocular muscle association is rarely described
	• Reflexes are characteristically absent; however, normal reflexes or hyperreflexia has been reported.
	✍ Neurological Equation Acutely progreesive parpapreis + arfelexia + minimal senspry symtomps and signs + Rasied CSF proetein =AMSAN
AMSAN (Acute motor-sensory axonal neuropathy)	• Chronic axonal polyneuropathy is a well known clinical complication of excessive alcohol consumption;
	• On the other hand, acute axonal polyneuropathy associated to alcohol abuse is less recognized.
Alcohol	
Vasculitis	
Toxins	
Thallium	
Arsenic	

Acute Dmeylinating Neuropathy

GBS (Guillain-Barré, syndrome)	▪ Clinically, acute motor axonal neuropathy presents as Guillain-Barré syndrome with an acute, ascending, and flaccid paralysis. ▪ This is distinguished from acute inflammatory demyelinating polyneuropathy primarily by electrophysiological studies. ▪ Diaphragmatic and cranial nerve weakness 50% ▪ Autonomic involvement > 50% ▪ Acute-to-subacute onset ▪ Nadir within 4 weeks ▪ Monophasic illness ▪ CELLS < 10 (Albumino-cytological ↳ Neurological Equation Acutely progreesive parpapreis + arfelexia + minimal senspry symtomps and signs + Rasied CSF proetein(after a week of onset) = GBS
HIV-AIDP	Peripheral neuropathies associated with infection HIV infection may be found in up to 50% of patients. ↳ Neurological Equation Known HIV+ + Acutely progreesive parpapreis + arfelexia + minimal senspry symtomps and signs + Rasied CSF proetein + raised CSF cell count > 10 (GBS will have CSF count of less then <10) = HIV-AIDP
Diphtheria	↳ Neurological Equation Descending paralysis + diplopa + dilated pupil + pharyngitis (travel to india as rare in Wetern world) = diphtheria

Dr Nadeem Akhtar

Subacute or chronic Demyelinating Neuropathy

CIDP	**Classic CIDP** CIDP is a chronic sensori - motor polyneuropahty characterised by • Progressive symmetrical weakness, sensory loss and reduced or absent deep tendon reflexes over period of more then two months • Clinical course can be relapsing and remitting, or steadily progressive.
Pearls	Disorders with distinct Prototype of this disorder characteristics, not likely to represent same disease as CIDP • Multifocal Motor Neuropathy • Anti MAG Neuropathy • PN associated with myeloma • PN associated with IgM MGUS • Chronic ataxic Neuropathy with opthalmoplegia (CANOMAD) ✍ Neurological Equation proximal onset + symmetrical distribution + Classical progression of more then 8 weeks + protein electrophoresis being normal (unlike MGUS and other paraprotein associated neuropathies = CIDP
MADSAM (Multifocal acquired demyelinating sensory and Motor neuropathy	Multifocal acquired demyelinating sensory and Motor (MADSAM) neuropathy is characterised by • Multifocal and asymmetric motor and sensory loss and conduction block • Other electrophysiological features of demyelination are also evident.

MMN (Multifocal motor neuropathy)	• Multifocal motor neuropathy characteristically causes asymmetrical weakness typically in distribution of peripheral nerves without sensory involvement.
	• High titirs of anti-GM1 antibodies are present in serum
Pearls	MMN has a clinical features of mononeuritis multiplex and neurophysiological evidence of persistent motor conduction block
	◆ CIDP, MADSAM neuropathy, MMN would show
	◆ electrophysiological features of demyelination such as
	◆ Conduction block
	◆ Slow conduction velocities
	◆ Temporal dispersion
	◆ Absent or prolonged F-waves
Pearl	MADSAM can be differentiated form MMN because it has sensory nerve involvement (MMN spares sensory nerves)
	And it is different form CIDP being asymmetric nerve involvement (CIDP results in symmetrical nerve involvement)
Moclonal gammopathy of unknown significance	Monoclonal proteins (IgM, IgG and IgA) in the serum and urine of patients with neuropathy may provide a marker for
	• Amyloidosis
	• Myeloma
	• Leukaemia
	• Waldenstrom's macroglobuliemia
	• MGUS

Diabetes	▪ Diabetes may predispose to CIDP
sHypothyroidism	▪ Neuropathy is not caused by direct infiltration but it is resulted due to the autoimmune phenomenon which Is not completely understood
Hodgkin lymphoma	▪ CSF shows mild lymphocytic pleocytosis and some increased gamma globulin.
HIV infection	
DADS Hereditary (CMT, etc) Medications Amiodarone	

Neuropathy with Autonomic implement

Diabetes mellitus:.	- Diabetic autonomic neuropathy can present with symptoms of postural hypotension, erectile dysfunction, diarrohea and sphincter disturbances - Cardiac denervation can also occur
Guillain-Barre syndrome:	- Autonomic involvement could presentas cardiac dysrhythmias (bradyarrhythmias, and tachyarrythmias) - Sometimes could be life-threatening and labile hypertension
Amyloidosis	- Clinically, it can present as cardiac or renal failure however peripheral neuropathy can be significant and in some cases can be a mode of presentation. - Confirming the diagnosis can be difficult and may need multiple diagnostic interventions including more than one biopsy - Immunohistochemical staining and genetic tests can help distinguishing between acquired and hereditaryforms of amyloidosis. - Neuropathy caused by amyloisdosis causes Pain in distal extremities loss of temperature sensation m trophic ulceration and autonomic signs and symptoms.
Porphyria	
vincristine	

Acute pure motor weakness

Guillain-Barré syndrome (GBS)

Vasculitic	<ul><li>Churg-Strauss Syndrome</li><li>Behcet`s Disease</li><li>Polarterits Nodosa (associated with hepatitis B or C infection)</li><li>Rheumatoid vasculitis</li><li>Sjogren's Syndrome</li><li>Wegner's Granulomatosis</li></ul>
Myasthenia gravis (acute presentions)	✍ Neurological Equation Acute generalised weakness (rare presentation of MG) + intact DTR (unlike GBS) + significant ptosis and diplopia + bulbar symptoms = Myasthenia gravis
Botulism	✍ Neurological Equation Descending paralysis +,+ptosis,+ diplopia, ophthalmoplegia, uncreative dialted pupils + bulbar paralysis = Botulism Alert : uncreative dilated pupils helps in distinguishing form other acute paralyses with ophthalmic involvement
Diphtheria	<ul><li>Inflammatory pharyngitis</li><li>Uncreativedilated pupil,</li><li>Ptosis,</li><li>Diplopia</li><li>Ophthalmoplegia,</li><li>bulbar paralysis along with descending paralysis</li></ul>
Lyme disease	
Tick Paralysis	<ul><li>History of outdoor activity</li></ul>

	<ul><li>Insect bites</li><li>Ascending paralysis that can progress and compromise bulbar and respiratory function</li></ul>Alert : History of outdoor activity and Insect bites
Porphyria	<ul><li>Colicky abdominal pain and</li><li>acute confusion and convulsions.</li><li>Associated weakness is due to a motor polyneuropathy strikingly more proximal than distal.</li><li>Ccan lead to complete flaccid quadriparesis with respiratory paralysis over a few days</li></ul>✍ Neurological Equation Episodes of confusion + abdominal pain + HTN (in young patient) + seizures (GTCs) + generalised weakness = AIP Alert : recurrent abdominal pain + Neurological symtopms
Periodic paralysis: familial	✍ Neurological Equation Family Hx of episodic weakness (duration could be varaible) + recurrent attacks + (on exposure of) carbohydrate or alcohol consumption + altered K serum level (can be high or low) = Periodic paralysis:

Neuropathy with rash

Brucella infection	▪ futures of neuropathy in patients from endemic areas and skin rash should alert this possibility.
Refsum`s disease	↝ Neurological Equation Predominalty sensory Sensory neuropathy + progressive visual loss + progressive deafness + skin rash(typical) = Refsums disease
Vascultiis	
Lyme Disease	

Predominantly Motor Neuropathy

GBS

Porphyria	Classically acute attacks compromise of symptoms including

- Colicky abdominal pain
- Cognitive symptoms (psychosis)
- Autonomic dysfunction- mainly due to vagal insufficiency.
- Peripheral neuropathy(clinical characteristic can resemble one of Guillain-Barre syndrome)

✍ Neurological Equation
Epiodes of confusion + abdominal pain + HTN (in young patient) + seziutres (GTCs) + genrlaise weaknesss = AIP

Lead, poising
- Ingestion of Asian folk medicines
- Paint Moonshine whiskey;
- Rum processed in lead pipes;
- Haematological findings include: anaemia, microcytic with basophilic stippling of erythrocytes.

✍ Neurological Equation
Parastheisa and suggestive of neuropathy+(suspiciousneurologist) + (being alert of possible sources of exposure) + anaemia + basophilic stippling = Lead, poisoning

Paraneoplastic
✍ Neurological Equation
known patient of Neoplastic disease ( ) + sensory symptoms + anti Hu antibodies + = Paraneoplastic neuropathy

Pearl
predominantly sensory neurapithy

Organophosphorus poisoning	<ul><li>Increased risk due to exposure (absorbed through skin , respiratory tract and Gi tract)</li><li>Petroleum additives</li><li>Flame retardants</li><li>Insecticides</li><li>Neuropathy would have</li><li>Motor: Distal weakness</li><li>Sensory: Distal loss & paresthesias</li><li>Neurological Equation Parenthesise + suspicious neurologist + being alert of possible sources of exposure + constricted pupil = = Organophosphorus poisoning</li></ul>
Isoniazid	<ul><li>Side effects including neuropathy are common in</li><li>Slow acetylators (metabolised slowly)</li><li>Neuroapthy is uncommon but</li><li>Classically is of large fibre, symmetrical and sensori-motor type</li></ul>

Generalized myasthenia

The conditions that can mimic conditions that mimic generalised myasthenia	
	▪ Generalised severe fatigue
	▪ Motor Neurone disease
	▪ Botulism
	▪ Lambert Eaton Myasthenia syndrome
	▪ Congenital myasthenia syndrome
	▪ Acquired Myasthenia-Penicillamine induced

Dr Nadeem Akhtar

Wrist drop

Radial Nerve involvement	▪ Penetrating trauma
	▪ Compressive.
	▪ use of crutches
	▪ drunken or drug-induced stupor with the arm over a chair or in wheelchair users
	▪ handcuffs and tight bracelets (The posterior interosseous syndrome)
Pearls	Can be due to any cause resulting in an injury anywhere along the nerve
	▪ lesions high in the axilla, can occur from improper use of crutches
	▪ when the spiral groove of the humerus is compressed on a hard wheelchair surface
	▪ Bilateral radial palsies suggest lead intoxication. Lead exposure may be occupational

Foot drop

Peroneal neuropathy	<ul><li>Peroneal neuropathyand L5 radiculopathy are by far the most common cause.</li><li>The Peroneal nerve is vulnerable to compression as it goes round the neck of the fibula</li><li>Damage can be a consequence of external trauma pressure or purely after a period of a prolonged bed rest in a immobile patient.</li><li>Weakness involves in the evertors of the foot and the dorsiflexors of the foot as well as toes</li></ul>Foot drop +(Ipanatar flexion+foot inversion spared) Numbness involving lower lateral and distal part of the leg= Superfical peroneal sensory involvement
Habitual Leg crossing	<ul><li>Habitual leg crossing while sitting is considered to be one of the most common cause and can improve when they avoid this habit</li></ul>
Other causes	<ul><li>Lumbosacral Plexopathy</li><li>Occupational eg. gardening</li><li>Weight loss</li><li>L5 radiculopathy</li><li>lumbar plexopathies</li><li>partial sciatic neuropathy</li><li>Multifocal motor neuropathy</li><li>Poliomyelitis</li><li>Parasagittal Tumors</li><li>Motor neuron disease</li></ul>

Dr Nadeem Akhtar

Neuropathy with autonomic involvement

diabetes mellitus	▪ Patients with diabetic autonomic neuropathy would have feature
	▪ Of postural hypotension, altered bowel mainly diarrhoea and sometime sphincter disturbances. Rhythm abnormalities due to cardiac denervation can occur.
Guillian-Barre Syndrome-	▪ Autonomic involvement is uncommon in GBS
	▪ it can cause cardiac dysrythhmia and labile hypertension
Amyloidosis	▪ Neuropathy caused by amyloisdosis causes
	▪ Pain in distal extremities
	▪ loss of temperature sensation
	▪ trophic ulceration and
	▪ autonomic signs and symptoms
Porphyria	
Vincristine	

Radial Neuropathy

Lesion In axilla	✍ Neurological Equations Wrist drop + triceps and brachioradialis reflexes are decreased= lesion is in the axilla (improper use of crutches)= Radial Nerve Plasy
Lesion In spiral groove	• The radial nerve is liable to damage as it winds round the spiral groove of the humerus • Triggering factors include blunt trauma and fractures of the humerus. • In `Saturday-night palsy, • Typically as a result of alcohol excess, the nerve is damaged during sleep in which the arm has been draped over the back of a chair. • The triceps is usually spared but the weakness is conspicuous in the wrist and finger extensors, brachioradialis and the supinator muscle. Sensory loss is often slight WRIST drop + Triceps reflex is preserved, + brachioradialis is decreased= lesion in spiral groove Triceps reflex is preserved +Brachioradialis reflex is intact = isolated posterior interosseous lesions (handcuffs and tight bracelets)

Dr Nadeem Akhtar

Rlapsing Neuropathy

GBS	
CIDP	
Mononeuritis multiplex	
Beriberi	Dry beriberi or acute nutritional neuropathy is uncommon in western world
	• Neuropathy can vae rapid course causing symmetrical weakness, paresthesia and neuropathic pains
	• This can be so acute that it can mimic Gullian-Barre syndrome
Refsums disease	Refsum`s Disease, a peroxisomal disorder
	Affecting metabolism of phytanic acid oxidation
	• Clinical features include
	• Retinitis pigmentosa
	• sensory neuropathy
	• Anosmia
	• deafness
	• ataxia
Porphyria	
Cauda equina lesión **Conus medullaris lesión**	• Cauda equine or conus medullaris lesion are common reasons of bilateral foot drop
	• Both of these syndrome have other features which include urinary retention and erectile dysfunction perianal and saddle anasthesia

	▪ CMS Conus medullaris lesions Knee jerks preserved but ankle jerks affected prapareisis is symmetrical + upgoing plantars
	▪ CES Cauda equina lesions Both ankle and knee jerks affected paprareis is assumetrical + downgoing platars
Distal myopathies, Welander myopathy	▪ Clinically, it may resemble (ALS) however ,Intervalbetween onset of symptoms diagnosis can be more than 15 years.
Hereditary neuropathy with liability to pressure palsy	▪ Classically motor (no sensory involvement) ▪ slowly progressive, ▪ asymmetric, ▪ distal weakness ▪ slow over years ▪ Neurophysiological studies would show conduction blocks present in two or more motor nerves(outside the possible entrapment sites) Ages Of 10 And 30 +Recurrent Focal Pressure+ Paslies, Of (Median, Ulnar, And Peroneal)Nerves +Improve In Hours To months + family history
Multifocal motor neuropathy with conduction block	
Bilateral sciatic neuropathy	
Bilateral Peroneal neuropathy	
Motor neuron disease	

Multifocal Neuropathies

MADSAM
Multifoal variant of CIDP
Hereditary neuropathy with liability to presuure palsy (HNPP)
Diabetes mellitus
Vasculitis
Leprosy
Sarcoidosis
Focal entrmpments
Myxedema
Rheumatoid artritis
Amyloidosis
Acromegaly
Neopalstic infiltración

Neuropathies with cranial nerve involvement

Diabetes Mellitus
Gullian-Barre syndrome
Lyme disese
Diphtheria
Sarcoidosis

Sensory Neuroapthy

Chronic gluten enteropathy
Crohn's disease

Hereditary sensory neuropathy
types I and IV

Carcinomatous sensory
neuronopathy

Paraneoplastic Paraproteinemias
Sjögren's syndrome

Idiopathic sensory neuronopathy
Primary biliary cirrhosis

The previous terms used
for these neuropathies
include peroneal muscular
atrophy and Charcot-
Marie-Tooth disease. They
are now classified on
the basis of histological
characteristics together
with the changes in
electrophysiological
function

Neuropathies with autonomic involvement

Pandysautonomia-	▪ Neuropathy may be accompanied by attacks of abdominal pain and psychiatric disturbance. ▪ Motor deficit predominates and is usually preceded by pain. autonomic dysfunction can lead to tachycardia and urinary retention
Diabetic neuropathy	▪ • Autonomic dysfunction can be marked in diabetic patients. ▪ • Clinical features will be impotence, bladder atony with incontinence, diarrhoea and also lead to disorder of blood pressure control
Porphyria Paraneoplastic neuropathy Amyloidosis Alcoholic neuropathy Arsenic, Thiamine deficiency Vincristine toxicity Guillain-Barré syndrome HIV/AIDS	

Dr Nadeem Akhtar

Headache - causes

Conditions- causing raised ICP and mimicking IIH

Diabetes Mellitus
Gullian-Barre syndrome
Lyme disese
Diphtheria
Sarcoidosis

- Addison Disease
- Addison Disease
- Hypoaparathyroidism
- COPD
- CCF
- CVST
- Jugular venous thrombosis
- Obstructive sleep aponea
- Medications
- Tetracyclines
- Vitamin-A and related compounds
- Anaboilic steroids
- Sudden withdrawal of corticosteroids

Pain sensitive structure

	Intracranial
	■ Periosteum
	■ Meningies
	■ Meningela aretreis
	■ Dural sinuses
	■ Proximal intracranl arteries
	■ Cranial nerves
	■ Thalamic nuclei
	■ Brian stem pain modulating cnetres
Red Flags in Hedache hisotry	■ New headache especially in over 50 y.o.
	■ Abrupt onset, unusually severe
	■ Change in usual headache pattern
	■ Associated with focal neurologic findings
	■ Change in LOC, personality, lethargy
	■ Fever, neck stiffness
	■ Systemic signs/symptoms
	■ Temporal artery tenderness

Thunderclap headache

Overview	A new abrupt onset peaking to be very intense within one minute of onset is called to be thunderclap headache
SAH	
RCVS (Reversible cerebral vasoconstriition Syndrome)	<ul><li>Recurrent headache is the only symptom in 75% of cases</li><li>Multiple over 1-4 week period is almost pathognomonic</li><li>Focal neurologic deficits and seizures in minority of patients</li><li>Alternating areas of constriction and dilatation – a.k.a. "beading" -- in several vascular territories</li><li>May be seen in large-to-medium-sized arteries of anterior or posterior circulation</li><li>Resolution within 3 months is most specific for RCVS</li></ul>
Spontaneous Intracranial Hypotension	<ul><li>Headache diffuse, more severe over frontal area +</li><li>Worst in upright position, especially 10+mins after standing up + better on lying down tinnitus(direct transmission CSF pressures)</li><li>LP would show an opening pressure below</li><li>Radiological features of low CSF pressure on MRI brain with contrast include pachymeningeal enhancement</li><li>Site of CSF leakage causing low CSF pressure can be found by conventional myelography or CT myelography</li></ul>

Dr Nadeem Akhtar

Cortical venous sinus thrombosis	▪ Patients younger then 40 years of age generally who have thrmbophillia or pregnant females or young females on hormonal contraception. Recent unusual headacheas subacute in onset + storke/TIA like symtomps(in abscnece of usual risk factors) + signs of intracranial hypertension+ CT brain evidence of infarcts (not confined to arterial vascular territory with heemrrhagic transformation))= CVST
Pituitary apoplexy	▪ A clinical syndrome due acute haemorrahge or infarction of the pituitary gland characterised by sudden headache, visual disturbance and possible collapse. ↳ Thunderclap headachee +vomiting + visual Disturbances(field Isos) + Opthalmoplegia –(CN III most common)+Meningismus+Decreased Consciousness= Pitutatry Apolpexy
Pearl	▪ CT will demonstrate pituitary mass but not sensitive in demonstrating haemorrhage or infarction (i.e. CT diagnostic in only 28 % of cases diagnostic in only 28% of cases, defined sellar amss in 72 % of cases ▪ MRI is the radiologic mode of choice (i.e. confirmed diagnosis in >90% of cases)
Intracranial haemorrhage	
Benign thunderclap headache (variant of migraine)	

Sleep Disorders - causes

Excessive sleepiness

Lack of sleep	• Significantly compromised sleep hygiene- inadequate time in bed
	• Sleep disruptions
	• Shift work
	• Obstructive sleep apnoea-OSA
	• Restless leg syndrome
	• Periodic Limb movement disorder
Narcolepsy	Classic tetrad Excessive daytime sleepiness+hypnogogic hallucinations(visual or auditory sensations while falling asleep or upon awakening) + sleep papralysis (unable to move or talk upon awakening)+ cataplexy (sudden loss of muscle control triggered by strong emotions such as laughing)
Neurological causes,	• Tumours of hypothalamus, pineal gland and brain stem
	• MS
	• Post head injury
	• Para median bilateral thalamic infarcts
	• head injury,
	• MS
Psychological	• Depression
	• Seasonal affective disorder
Psychological	
Idiopathic hypersomnolence	
Drugs	

Seizure Disorders - causes

Causes of seizures with Encephalopathy

Corticovensou thrombosis

Toxic or metabolic enecpahlopathy

Venous thrombosis

Demyelianting disorders (ADEM)

PRES

Cerebral vascultis

Top of basilar syndrome
Encephalitis

Dementia and related Disorders

Treatable causes of Dementia

NPH
Vitamin B12 Deficiency
Hypothyroidism
Syphillis
Depression (pseudo-dementia)
SDH (subdural hematoma)

Common causes Dementia in elderly

Alzheimer's	<ul><li>Most common cause of dementia in people over 65 years of agethe age of 65 and about 50% of people over the age of 85 years are affected.</li><li>Clinical features include:</li><li>Memory loss (mainly long-term)</li><li>Language problems</li><li>Mood swings</li><li>Personality changes</li></ul>
Vascular	<ul><li>History and evidence of focal neurological symptoms</li><li>Cognitive deficits</li><li>Impaired memory</li><li>Apahsia, apraixa, agnosia</li><li>Impaired social and occupational function</li><li>Focal neurological symptoms and signs or evidence of cerebro-vascular disease</li></ul>

Dementia with Lewy Bodies(DLB)	Third most common cause of dementia
	- Cognitive compromise is normally a presenting feature
	- Minority present with parkinsonism
	- Some with psychiatric disorder with dementia
	- persistent well-formed visual hallucinations,
	- Others with orthostatic hypotension, falls or transient disturbances of consciousness
	- Sporadic (rarely familial)
	- Generally duration of Lewy body dementiais often shorter than many other neurodegenrative dirorders
	✎ Neurological Equation Central feature is progressive cognitive decline + Pronounced fluctuations + Recurrent visual hallucinations + Parkinsonism- DLB
Pearl	*Synucliopathies (Abnormal aggregation of proteins, including, alpha-synuclein, neurofilament and ubiquitin)* *PD* *Dementia with Lewy bodies* *MSA* *Amyotrophic lateral sclerosis* *Hallervorden-Spatz syndrome*

Rapidly progressing Dementias

Jakob-Creutzfeldt disease (CJD; sporadic, iatrogenic, familial)

- Creutzfeldt-Jakob disease (CJD) generally is rapidly progressive and invariably fatal neurodegenerative disorder due to abnormal accumulation of intracellular prion protein

- Cognitive compromise and higher mental function deficits is main clinical feature

- These deficits progress over few weeks or months to a state of a profound dementia

- Some cases can present with visual impairment , coordination deficits or cerebellar signs

- Most patients (about 90%) with CJD has myoclonus at some stage in illness.

- Startle myoclonus is also frequent

Pearl

Dementia with myoclonus can also be due to

- dementia with Lewy bodies,

- corticobasal degeneration,

- cryptococcal encephalitis,

- Unverricht-Lundborg disease.

- Alzheimer's disease (AD),

Frontotemporal dementia (FTD)	• Accounts for up to 3-20% of dementia • Mean age of onset is 53 years (this makes FTD a common cause of dementia in younger population) • It is considered to be third behind Alzheimer's dieses and Lewy body dementia **Core Features** • Insidious onset and relatively slow progression • Early decline of interpersonal and social conduct • Emotional blunting **Supportive features:** • Distractibility • Mental rigidity • Perseverative behaviour • Hyperoarlaity **MRI** • Prominent frontal and temporal lobar atrophy
Corticobasal degeneration (CBD)	Classical features clinical hallmark are • Asymmetric parkinsonism (including tremors) • Limb Dystonia • Myoclonus • Alien limb phenomenon • Gait abnormalities • Cortical sensory loss • Apraxia

Dr Nadeem Akhtar

Alzheimer's disease (AD)	• Many features of CBD, including myoclonus, alien limb and visual, sensory and motor deficit
Vascular	
Toxic-Metabolic	
Autoimmune	

Dementia with Extrapyramidal signs

Progressive Supranuclear Palsy (PSP)	• Patients with PSP can develop dementia • With other features of akinetic-rigid parkinsonism syndrome Classically, earlier in disease they have : • axial rigidity • Postural instability • Further progression causes bulbar involvement (swallowing and speech disturbances) and eventually hypokinetic mute state
PD Dementia	
Dementia with Lewy bodies	
MSA	
Amyotrophic lateral sclerosis	
Hallervorden-Spatz syndrome	

Cranial Nerve and related - causes

Pupillary Abnormality

Pupils	<ul><li>Their size vary from 1 to 8mm in diameter</li><li>Normal pupils range from 3 to 5mm in ambient light conditions</li><li>Miotic pupils are less than 3mm</li><li>Mydriatic pupils are greater than 7mm</li></ul>
Abnormally Large Pupil **CN III Palsy**	
Acute Adie's tonic pupil	<ul><li>Damage to ciliary ganglion or post gangionic fibers of short ciliary nerve(parasympathetic pathway lesion)</li></ul> ✍ Neurological equation Females + The affected eye is dilated and reacts poorly to light (poor direct and consensual response) + Near reaction is strong, slow, and tonic
Pearl	<ul><li>The diagnosis is made by demonstrating reduced deep tendon reflexes (mainly knee and ankles)</li><li>Some of these patients have patch hypohydrosis due to involvement of sudomotor fibres called Ross Syndrome</li></ul>
Pearl	<ul><li>1% pilocarpine test will constrict a compressive or tonic pupil but not a pharmacological one</li></ul>
Benign Episodic Pupillary Mydriasis	<ul><li>In women who suffer migraines:</li><li>lasts minutes to one week but usually about 12 hours.</li><li>May or may not react to light</li></ul>
Drugs	
Iris damage	

Abnormally small pupil	
Horner`s Syndrome	ptosis + miosis, facial anhydrosis (sweat gland denervation) iris heterochromia (congenital Horner's) + Pupil reacts normally to light and near Reflex
	• Bilateral, asymmetrically miotic pupils which are irregular
	• Poor dilation with poor response to light but brisk near response
	• Hallmark of tertiary neurosyphilis
Argyll- Robertson Pupil	
Pearl	• Total blindness due to bilateral cortical lesion does not affect the light reflex
	• Affected pupils are larger , often light / near dissociation like A-R pupils
	Parinaud's syndrome:
	• Nystagmus on attempted convergence and retraction on up gaze with decreased accommodation
	• Can be caused by a pinealoma, stroke, or MS
Simple anisocoria	
Drugs	

Ptosis - causes

Horner's syndrome	▪ Ptosis - denervation of Müller muscle + "Reverse ptosis" - lower lid elevation + Miosis - greater in dim light (dilation lag) + Anhidrosis +impaired flushing and sweating (In First-order: ipsilateral body in case of Second-order: ipsilateral face , in the case of Post-ganglionic (third-order): absent or limited)
Pearls	▪ Iris heterochromia – affected iris is lighter Congenital or children < 2 yrs
	▪ Long-standing lesions
First-order neuron lesion	(hypothalamus to C8-T2)
	▪ Cerebral vascular accident
	▪ lateral medullary syndrome
	▪ Arnold-Chiari malformation
	▪ Basal meningitis
	▪ Intrapontine hemorrhage
	▪ Demyelinating disease (eg, multiple sclerosis)
	▪ Traumatic dissection of the vertebral artery
Second –order neron disorder	▪ Pancoast tumor
	▪ Cervical Rib
	▪ Aneurysm/Dissection of arota
	▪ Central venous catheterisation
	▪ Birth Trauma or injury to lower brachial plexus
Third order neuron lesions	preganglionic lesions (T1 to SCG)
	▪ Internal carotid artery dissection
	▪ Carotid cavernous fistula
	▪ Cluster or migraine headaches

Pearl	<ul><li>Second order neuron runs along the surface of the lung, can be affected by a Pancoast tumor</li><li>Third order neuron runs with the carotid artery then with the ophthalmic division of cranial nerve V</li></ul>

Visual Loss Causes
Sudden visual loss

Seconds	Transient visual obscurations (TVOs) precipitated by postural changes or straining ▪ Chronic papilledema due to increased ICP ▪ Hypotension and hypoperfusion
Minutes	Monocular transient visual loss: ▪ Amarausus fujax) is due to emboli from carotid vessels or heart. Sudden onset lasting for 5-15 minutes. ▪ It is described as acurtain being pulled downwards in fornt of the eye ▪ Loss may be quadrantic or total monocular loss
Pearl	TVOs that are gaze-evoked suggest orbital tumors
Minutes to an hour	Migraine ▪ Retinal migraine is rare and results from transient vasospasm that generally responds to calcium channel blockers ▪ Closed angle glaucoma ▪ Accompanied by halos around lights and may not always be associated with redness and pain
PERSISTENT (>24 hr)	
Sudden, Painless	▪ Optic neuritis/neuropathy ▪ Temporal arteritis ▪ Anterior ischameic optic neuropathy (AION) ▪ Central retinal vein occlusion (CRVO)

Pearl	• Rapid onset is typical of demyelinating, inflmatory, ischameic and traumatic causes • An insidious and slow clinical course hints towards compressive toxic nutritional and hereditary causes. • The classic signs of optic neuropathy and visual defects and abnormal papillary response
Anterior ischameic optic neuropathy (AION)	A common cause of sudden monocular blindness, particularly in elderly individuals, is ischemic optic neuropathy. • It is because of involvement of ciliary vessels that supply the optic nerve. • Clinical features strongly suggestive of artertic ischameic optic neuropathy due to giant cell artertis (GCA) are: • Visual loss sometimes associated • Temporal area pain and tenderness • Jaw claudication • Constitutional symptoms including weight loss, jaw claudication and myalgias
Sudden, Painful	• Demyelination- most common • Lyme disease
Retrobulbar neuritis Papillitis	• Sinus-related (ethmoiditis) • Viral infections and immunization in children (bilateral)
Neuroretinitis	• Demyelination (uncommon • Syphilis
Close angle glucoma	• Cat-scratch fever

Gradual and painless	
	▪ Age related
	▪ Macular degeneration
	▪ Chronic retinal disease
	▪ CMV retinopathy
	▪ Diabetic retinopathy
	▪ Open-angle glaucoma
	▪ Cataract
Causes of episodic visual loss	▪ Migraine
	▪ Antiphospholipid antibody syndrome and systemic lupus erythematosus
	▪ Early tumor compression of the optic nerve
	▪ Takayasu aortic arteritis
	▪ Viral neuroretinitis
	▪ Carotid stenosis or dissection
	▪ Embolism to the retina Intrinsic central retinal artery atherosclerotic disease
Causes of anosmia	▪ Smoking
	▪ Chronic
	▪ Overuse of nasal vasoconstrictors
	▪ Cranial surgery
	▪ Subarachnoid hemorrhage,
	▪ meningitis
	▪ Toxic (-aminoglycosides, tetracyclines, corticosteroids, methotrexate, opiates, L-dopa)
	▪ Metabolic (thiamine deficiency,
	▪ adrenal and thyroid deficiency
	▪ Wegener ganulomatosis
	▪ Central Degenerative diseases (Parkinson, Alzheimer, Huntington)
	▪ Temporal lobe epilepsy
	▪ Malingering and hysteria

Facial pain

Trigeminal Neurlagia **Trigeminal neuralgia**	Characterized by recurring paroxysmal • Brief duration (for seconds) • Intense pain • In distribution of trigeminal nerve • Triggered by chewing, talking or touching the affected side of face
Glosopharyngeal Neuralgia	• Episodes of sudden pain in the region of tonsil on one side only • Pain is intense and could last for 1-2 hours and generally recurrent on daily basis
Sluder`s and Vidian Neuralgia	• Severe pain in and around nose, eyes , cheek and lower jaw • Could be due to lesion in sphenopalatine ganglion or vidian nerve
Post traumatic Neurlagia	• Pareitla and occipital region • Good prognosis and most recover
Neuroma	
Atypical facial pain	• Pain felt over the cheek, nose upper lip or jaw • Symmerticla and bialterla • Burning , shooting and Aching • Associated with reddening of the skin • May last for hours days or weeks • Lasts for hours, days or weeks
Post herpetic Neurlagia	• Herpes Zoster can affect trigeminal nerve ganglion • Rash which is vesicular covers one division which is commonly the opthamic division

Dr Nadeem Akhtar

Central lesion	▪ Rarely MS, tumors of brian stem, thrombotic lessiosn, metsataiss occult naspharyngela ca
Temporomandibular joint pain	

Cranial nerve palsies

3rd cranial nerve palsies	*Pupil spared* • If the pupil is spared but all the muscles are involved then cause • is like to be ischaemic. Diabetic neuropathy is a common case. • Demyelinating cause is uncommon. *If the pupil is involved* Common causes are • PCOM Aneurysm • If pupils is dilated and fixed – it suggest 3rd nerve external compression
	• Trauma • Intracranial mass lesion
Pearls	• it is rare (5%) and uncommon to have pupillary sparing in PCOM aneurysm. • Complete unilateral 3rd Nerve palsy In increasingly unresponsive patient suggest transtentorial herniation.
4th	• Closed head trauma without skull fracture is relatively common causes of unilateral and bilateral palsies • Aneurysm, tumors and multiple sclerosis are rare causes
5th	• Idiopathic • Idiopathic cases are common • Generally no other cranial nerves are involved and improvement generally occurs in about two months • Neoplasm • Vascular Malformation of the brain stem

Dr Nadeem Akhtar

	• Vascular insult
	• Multiple sclerosis
	• Sjogren syndrome
	• RA
6th	• Diabetes
	• Diabetic infarction is one of the more common causes
	• Trauma
	• Basal Meningitis
	• Carcinmaotus Meningitis
	• Wernicke's encephalopathy
	• Multiple sclerosis
	• Inflammatory, infectious and Neopalstic infiltrates entrapping 6th nerve
7th Nerve Bell's Palsy	• Complete weakness of the entire half of the face differentiates Bell's Palsy from supranuclear lesions (Stroke, cerebral tumors) in which the paralysis and weakness involves the lower half of the face while sparing frontalis and orbicularis
Ramsay Hunt's Syndrome Middle ear or mastoid infections Sarcoidosis Lyme disease	• Chronic meningitis • CP angle tumor • Glomus Jugulare tumors • CVA
8th Nerve Palsy 8th Nerve Palsy	• Multiple sclerosis • Syringomyelia • vascultiits

Ophthalmoplegia causes

Complex ophthalmoplegia

Acute cases **Inflammatory**	Tolosa –Hunt Syndrome • it is caused by granulmatosis • inflammation of the cavernous sinus or superior orbital fissure • Severe periorbital or retro-orbital pain of acute or subacute onset is typical • Complex ophthalmopareis is important clinical feature • Ptosis and papillary dysfunction can occur • involvement of ophthalmic division of Trigeminal nerve
Wernickes Wernickes	
Miller Fischer syndrome	
Myasthenia gravis	
Ocular	
Generalised	
ophthalmoplegic migraine	• This clinical syndrome is characterized by paroxysmal, recurrent, • transient partial or total ophthalmoplegia • associated with severe hemicrania and with or without scintillating scotoma, homonymous hemianopsia, depression, general malaise, nausea and • It frequently progresses to total and permanent unilateral oculomotor paralysis

Pearl	There is a history of typical migraine attacks, without ocular muscle paralysis, of many years' duration, prior to the first attack of ophthalmoplegia. The individual attack may begin, as in any other form of migraine, with scintillating scotoma and homonymous hemianopsia, followed by severe
Chronic causes	
Oculopharyngeal dystrophy	
Infectious aetiologies	
Micro -vascul Neoplasm	
orbital pseudotumor	
giant cell arteritis	
sarcoidosis	

Ptosis

Congenital	• Developmental dystrophy of levator muscle • Occasionally associated with weakness of superior rectus • Congenital Myasthenia Blepharophimosis syndrome • Rare congenital Dominant inheritance + Moderate to severe symmetrical ptosis + • Poorly developed nasal bridge • + Telecanthus (lateral displacement of medial canthus
Acquired Neurogenic	• Third nerve paralysis or due to reduced sympathetic innervations • Horner syndrome – ptosis, anhydrosis and miosis Mysthenia gravis • bilateral ptosis, increases by prolonged fixation or attempt to look up , external ophthalmoplegia partial or complete Conformation by prostigmin or edrophonium injection test

Myogenic	▪ Myotonic dystrophy ▪ Chronic progressive exophthalmoplegia ▪ Occulopharyngeal dystrophy ▪ Aponeurotic Ptosis *CPEO* ▪ Ptosis - slowly progressive and Symmetrical + Ophthalmoplegia – slowlyprogressive and symmetrical (no diplopia)
Aponeurotic Ptosis	Aponeurotic Ptosis ▪ Is involutional is due to weakness or disinsertion of LPS aponeurosis from ant surface of tarsal plate
Mechanical	
Pearls	Normal position of lids (Margin Reflex Distance (MRD)- ▪ Normal MRD is 4 mm +/- 1 mm ▪ Ptosis of less than 2 mm – Mild ▪ Ptosis of 3 mm – moderate ▪ Ptosis of 4 mm or more – severe
Vascular	Anterior ischaemic optic neuropathy (AION) Iscahmic central retinal vein occlusion (CRVO)
Inflammatory.	Optic neuritis
Infiltrative, compressive, inherited, nutritional.	Optic neuropathy. Compressive Toxic Metabolic optic neuropathy

Dr Nadeem Akhtar

Encephalitis - causes

Encephalitis

Herpes simplex virus **Type-1** **Type- 2**	This is most common cause of acute viral infective encephalitis 10-20 % of all cases Type-1 virus casuses more then 95% of cases of HSV encephalitis MRI brain shows – lesions in temporal lobe in over 80 % of cases and frontal lobe in about 36 % of cases.
Varicella-zoster virus **Hepatitis B virus** **Epstein-Barr virus** **Cytomegalovirus** **Flaviviridae: West Nile,** **Japanese, Dengue** **Poliovirus** **Rabies** **Meales, mumps** **Influenza** **(postinfectious?)** **HIV**	
Non-Viral causes of acute **encephalitis/myelitis**	<ul><li>Rickettsia,</li><li>Brucella,</li><li>Mycoplasma,</li><li>Bartonella,</li><li>Lyme dis</li><li>syphilis</li><li>Toxoplasma,</li><li>Plasmodium falciparum</li><li>Vasculitis</li><li>Carcinoma</li><li>Drug reaction (chemotherapy: methotrexate</li></ul>

ADEM	<ul><li>This is an inflammatory demyelinating disorder of sub cortical white matter most frequently seen in children</li><li>often evolving from antecedent infection or immunization.</li><li>Typical presentation: encephalitic signs with non specific CSF changes and minimal or no changes on CT brain MRI- gold standard for diagnosis T2 weighted images show areas of prolonged T2 in subcortical white matter, usually assymetrical.</li></ul>
Autoimmune Encephalitis	
VGKC Antibodies	
NMDA Antibodies	

Vertigo - causes

Vertigo

	Defined as an illusion or hallucination of movement.
	Both vertigo and disequilibrium imply a loss of balance but vertigo involves a sense of motion. Patient feels if he is turning in stationary environment or objects around him or her moving.
	Other description • Rocking tilting • Somersaulting • Descending in elevator
Central	
Benign paroxysmal positional vertigo (BPPV) (50%)	Otoliths become detached form hair cells in uticle inappropriately enter the posterior semicircular • On turning head • After a few seconds delay , vertigo occurs • Resolves within a minute if patient does not move
Vestibular neuritis	Characterized by triad of : • Vertigo • Tinnitus • Hearing loss (sesnorineural) Chronic relapsing illness (? familial) Due to a build-up of endolymphatic pressure in the labyrinth. Treatment: vestibular suppressants

Vestibular migraine
Senile vestibulopathy
Meniere's disese
Labyrinthitis (suppurative,
serous, toxic, chronic)
FB in ear canal
Acute otitis media
Perilymphatic fistula
Toxic/ Drug induced

- Alcohol

- Aminoglycosides

- Chloramphenicol

- Minocycline

- Quinine

Neuroemergencies

Cortical Venous Thrombosis

Overview	Disease characterized clinically by headache, papilledema, seizures, focal deficits and if not treated in time then progress to coma
	Involvement of sinuses in order of frequency
	<ul><li>Superior sagittal sinus</li><li>Lateral sinus<ul><li>Right</li><li>Left</li><li>Both</li></ul></li><li>Straight sinus</li><li>Cavernous sinus</li><li>Cerebral veins</li><li>Superficial</li><li>Deep</li></ul>
Risk factors	Idioapthic Local: Head injury Infective- intracranial regional infection Neurosurgery Tumors Infusions into jugular vein General:<ul><li>Postoperative,</li><li>pregnancy/postpartum,</li><li>inflammatory bowel disease,</li><li>malignancy,</li><li>thrombophilia</li></ul>— 5% CVT have a detectable thrombophilia (APC resistance; antithrombin, protein C or S deficeincy, antiphospholipid syndrome)

Pearl	20% of cases of CVST in women < 50 years old
Clinical Features	**Headache with focal signs** - 90% of adults of often builds up progressively but can imitate a subarachnoid hemorrhage. **Behavioral symptoms / encephalopathy** - delirium, amnesia, mutism - altered consciouisness from isolated behavioral changes to coma. **Seizure** seizure-can occur upto 40% patients (much more common compared to other types of stroke) **Stroke** - signs and symptoms of stroke initially would start form on hemisphere then it can progress to involve the other hemisphere within few days. - Cortical lesion can occur on both sides of superior sagittal sinus pearl combination of headache, **IIH** - Can present as isolated IIH (40% of cases)
Pearl	seizure and focal neurological deficits is strongly indicative of CVST

Investigations	**Thrombophillia screen:**
	▪ Exclude possibilities for thrombohila if patient is young with family history of hand recurrence.
	▪ Factor V Leiden mutation, prothrombigene mutation, protein C deficiency, protein is deficiency, anti-thrombi in III deficiency.
	CT Head
	◆ Infarction in nonarterial distribution (often hemorrhagic)
	◆ Empty delta sign
	◆ Dense triangle sign
	◆ Cord sign
	MRI/V
	◆ **Early:** absence of flow void & isointense on T1 for occluded vessel; Hypointense on T2
	◆ **Late:** hyperuintense thrombus on T1 & T2
Management	
Anticoagulation rationale	▪ Arrest the thrombotic process.
	▪ Tendency for venous infarcts to become hemorrhagic.
	▪ 40% of patients with sinus thrombosis – hemorrhagic infact prior to anticoagulation.
	▪ Anticoagulation is safe, even in the setting of ICH.
Heparin or cleaxane, which one to use	▪ No data comparing the effect of Unfractionated Heparin with Low molecular weight heparin. Generally heparin is preferred over cleaxne.

When to start warfarin and for how long?	<ul><li>Best possible duration of anticoagulation treatment after the acute phase is unidentified.</li><li>3-6 months is typical, especially if CVST was due to a transient risk factor Recurrent Sinus thrombosis 2% of patients 80% of relapses occurred within first 2 years conversion to warfarin (INR goal, 2.0–3.0) once the patient is stable.</li></ul>
Is there any place for Thrombolysis	<ul><li>Variable results and insufficient data available should be limited to patients with a poor prognosis, in centers where the staff have experience in interventional radiology.</li><li>Data limited to case reports consider if clinical deterioration despite adequate anticoagulation.</li></ul>
Discontinue	<ul><li>Discontinuation of oral contraceptives, hormone replacement, and other prothromobtic drugs rule out other cause.</li></ul>
Intracranial Hypertension	<ul><li>LP if not contraindicated – measure CSF pressure aim to lower ICP, relieve the headache, reduce papilloedema.</li><li>Oral acetazolemide.</li><li>If repeated LP and oral acetazolemide do not control the ICP within 2 weeks, surgical drainage is indicated, usually by a lumboperitoneal shunt.</li></ul>

Dr Nadeem Akhtar

Management in Pregnancy	• Management of obstetric CVT is not different from that of CVT unrelated to pregnancy.
	• Hence it includes supportive care, seizure control, measures to lower intracranial pressure, search and treatment of possible infection.
	• To prevent further thrombosis, anticoagulation is the preferred treatment.
	✎ Neurological equation Recent unusual headache stroke like symptoms in the + absence of usual risk factors + intracranial hypertension + CT evidence of hemorrhagic infarcts, especially if not confined to arterial vascular territories.
Complication	• Recurrent Sinus thrombosis 2% of patients (ISCVT)
	• 80% of relapses occurred within first 2 years, mean latency of 10.3 months (Mehraein)
	• Extra cranial thrombotic event within one year - 4%
Pearls	• All included patients who had hemorrhagic infarcts prior to treatment, no increased or new cerebral hemorrhages developed after treatment with heparin.

Prognosis	<ul><li>Very few patients dependent at 18/12.</li><li>Death / dependency 13 %.</li><li>Complete recovery 77%.</li><li>Contrast with arterial stroke – proportion of permanently dependent patient ranges between 1-2/3 of survivors</li></ul>

Dr Nadeem Akhtar

Vertebral Artery Dissection

Overview	Dissection is due to tear in the intima of vertebral artery Intracrianl arery dissection is mostly spontaneous

- Peak incidence 40's
- 2.5% of first strokes
- Carotid Dissection, males = females
- Vertebral Dissection - females > males
- VAD is responsible
- for up to 2.5 % of CVA
- up to 5 % of strokes up to in patients below 45 years of age
- About 15 % of lower brian stem stroke

Risk factors

- Spontaneous (most commony)
- Cervical manipulation
- Nose blowing
- Minor neck trauma
- Ceiling manipulation
- Judo
- Marfan Syndrome
- Ehlers Danlos Syndrome

Clinical Features

Headache

- Pain- mainly in occipital distribution as well in neck in 50-70 % of cases

TIA (posterior circulation)

- precedes the stroke in about 10 % of cases
- latency can be up to 2 weeks

	Asymptotomatic
	◆ about 5-8%
	Vertebrobasilar circulation ischaemia.
	◆ Present with posterior circulation and infarctions or SAH
	◆ Upto 40% of posterior circulation stroke are due to vertebral artery dissection
Signs	**Signs of Lateral medullary syndrome including**
	◆ Ataxia
	◆ Nystagmus
	◆ Ipsilateral Horner syndrome
	◆ Ispilateral-imapiered fine touch and proprioception
	◆ Contralateral – impaired pain and temperature sensation
	◆ Lower cranial nerve involvement
	Intracranial vertebral dissection
	◆ Brian stem infarct
	◆ SAH (due to subadvedntial or transmural dissection)
	◆ Mass effect on lower cranial nerves
Investigations	▪ Ct Head
	◆ To rule out complication of subarachnoid hemorrhage. (Once ruled out it is possible to start anticoagulant
	▪ Magnetic resonance angiography (MRA)
	◆ MRA - is non-invasive investigation of choice.

Dr Nadeem Akhtar

	◆ In many cases MRI/MRA is sufficient to diagnose Sensitivity of MRA is similar to CTA
	▪ Angiography
Management	
Thrombolysis	Acute stroke - thrombolysis
Anticoagulant therapy	Anticoagulant therapy (IV heparin or unfratinoated Heparin, eg Deltaparin in patients without any haemorrhage Two options
	▪ Intravenous Heparin- in standard dose (Generally starting dose is 1000 units) should be started and titrated according to APTT ratio.
	▪ Deltaparin- is alternative option and should be used in therapeutic dosage.
	▪ Once patient is clinically stable then it should be switched to oral warfarin. Total duration is 3-6 moths
Re-imaging	▪ Some physician consider reimaging, (Doppler or MRA) at 3- to 6 months to see recanalisation before stopping oral anticoagulation. Otherwise, anticoagulation would continue for six months

Treatment in tricky situations

Patients with hemorrhagic transformation or large infarct	▪ Instead of anticoagulation these patients cna be treated with 100 to 300 mg per day of oral aspirin
Patients with intracranial vertebral artery dissection	▪ Patients with intracranial sVAD or intracranial extension of sVAD were treated with 300mg aspirin per day for 3-6 months
Pearl	Role of anticoagulation is not in intracranial dissection is not well established
Prognosis	▪ Vertebral artery dissection generally carries a benign prognosis Ischemic stroke generally tend to occur within a month particularly in first week ▪ Recnanlisation occurs within few weeks or months ▪ Recurrence of dissection risk is about 1% per year
Pearl	Bad prognostic markers are, extension into basilar artery and or associated SAH
Pearls	▪ Up to 40% of posterior circulation stroke are caused by vertebral artery dissection ▪ Fibromuscular dysplasia was found to be the cause of dissection in 11% cases who underwent DSA

Carotid Artery Dissection

Overview	Hemorrhage within the carotid wall Splitting along natural lines of cleavage' Thrombus (blood clot) can build at the site of the tear Thrombus can break off and be swept upstream
	Association with arteriopathy/trauma
	<ul><li>75% cases extracranial internal carotid artery (M=F)</li><li>15% extracranial vertebral artery (F>M)</li><li>remainder intracranial internal carotid/vertebral, middle cerebral artery or basilar</li><li>15% of cases are bilateral</li><li>ntracranial dissection more common in male adolescents / <30 years Extracranial carotid artery = mean age of 40 years.</li></ul>
Risk factors/ causes	<ul><li>Spontaneous</li><li>Traumatic, e.g. blunt carotid injury, penetrating injury ?</li><li>minor trauma ~25%</li><li>Chiropractic Manipulation</li><li>Connective Tissue Disorder, (0-18% in case series - ie most do not) e.g.</li></ul>
	Connective Tissue Disorder, (0-18% in case series - ie most do not) e.g. Marfan's, Ehrlos Danlos Type IV, FMD, ADPCKD,

Clinical Features	Extra-cranial carotid artery dissection Classic triad includes - Ipsilateral neck pain or headache - Partial Horner Syndrome (interruption of postganglionic sympathetic fibres resulting in partial ptosis and miosis) - Amurosis Fugax (transient, short lasting loss of vision due to compromised retinal flow) - Ipsilateral neck swelling (rare) - Tinnitus- ipsilateral can occur upto 25% of cases) - Focal weakness.
Pearl	Seat belt sign- redness and echymosis in the distribution of neck and chest.
Investigations **Imaging for Dissection**	- Angiogram – gold standard (intimal flap / double barrel lumen may be merely irregular vessel wall). - CTA – seeing similar results to angiogram. - MRA- Not very sensitive modality, may fail to detect intramural hematoma wiithin first 24-48 hours of dissection. Signs on MRA include Filling defect Polo mint sign Changes in the calibre and vessel marginirregularities
Imaging for Ischemia	- MRI Brian with DWI
Management **Specific Measures**	
Thrombolysis for occlusion	- Thrombolysis for occlusion

Dr Nadeem Akhtar

Anticoagulation	▪ Secondary stroke prevention with anticoagulation
	▪ Within 7 days of onset of symptoms, treat for at least 3 months
	▪ Treat for at least 3 months (tend to 6)
follow-up CT-angiogram	▪ A follow up CT-angiogram, Duplex ultrasound should be reapeted few months after the intial event to revaluate the dissection.
Neurosurgical interventions	▪ Arterialreconstruction/repairviaopen surgery, stenting and/or coiling of stenosis or flap or pseudoaneurysm
Prognosis	▪ Mortality 5-1% but >70% for intracranial
Pearls	Peak incidence 40's
	2.5% of first strokes
	Carotid Disection incidence, males = females
	Vertebral dissection incidence, females > males

Neurolpetic Malignant Syndrome

Overview/criteria	<ul><li>Incidence 1% (0.02–3.23)</li><li>Pre-NMS<ul><li>psychomotor agitation</li><li>dehydration</li></ul></li><li>Recent or current therapy with dopamine blocking drug mainly neuroleptics</li><li>Recently stopped a dopamine agonist eg L-dopa</li></ul>
Risk factors	<ul><li>neuroleptics</li><li>other drugs eg metoclopramide</li><li>Dehydration</li><li>Agitation</li><li>Malnutrition</li></ul>
Clinical Features	Major clinical features (all generally occur within 24 h)<ul><li>fever > 37.5oC (no other cause)</li><li>extrapyramidal features</li><li>autonomic dysfunction<ul><li>tachycardia</li><li>diaphoresis</li><li>incontinence</li><li>hypertension</li><li>systolic > 30 mmHg above baseline or</li><li>diastolic > 20 mmHg above baseline</li><li>labile BP variability of > 30 mmHg systolic or >20 mmHg diastolic between readings</li></ul></li></ul>

Dr Nadeem Akhtar

Investigations

- Rise in creatinine kinase

- leucocytosis > 15,000x109/L

- Help confirm diagnosis therapeutic response to dopamine agonist

Management

General Measures	<ul><li>temperature tds</li><li>monitor blood pressure tds</li><li>record episodes of diaphoresis</li><li>On suspicion assess for other medical illness FBC, MBA, CK, serum iron</li><li>withdraw all dopamine blocking drugs</li></ul>
Specific Measures	<ul><li>Bromocriptine 7.5 to 45 mg/day in three divided doses- Reponse should be monitored clinically as well by CK levels. Improvement is expected in 24- to 48 Hours</li><li>Continue for upto two weeks then taper over 1-2 weeks,</li><li>May need to continue longer if there is a depot preparations)</li><li>Dantrolene: 2-3 mg/kg extreme rigidity, very high fever (> 400C), unable to tolerate oral treatment.</li></ul>
Prognosis	<ul><li>NMS can sometimes be prolonged. If neuroleptics need to be restarted then it should be given at low initial doses and preference should be given to atypical neuroleptics.</li></ul>

Dr Nadeem Akhtar

Posterior Reversible Encephalopathy syndrome

Overview	Posterior reversible encephalopathy syndrome is a clinco-radiologic entity. The radiological findings are symmetrical and predominate in posterior white matter. Typically it is reversible if insult is removed. However, this term is challenged as it not always posterior and not always reversible, as there is mortality in up-to 15 % of cases.

Risk factors	PRES is associated with vast majority of conditions
	
	- HTN/hypertensive Encephalopathy (61%)
	- Renal failure with HTN
	- Drugs (19 %)
	  - Immunosuppressive and chemotherapeutics agents - (alkylating agents, Anti TNF, Anti-metabolite, mitotic inhibiters)
	  - Antidepressants (MAO inhibitors tricyclic antidepressants)
	  - Pre-eclampsia/eclampsia (more common in women, even when eclampsia excluded)
	- IV Caffeine
	- Recreational (cocaine, LSD, amphetamines, PCP)
	- Blood transfusions
	- Collagen vascular disease
	- Sepsis

Clinical Features	**Headache** • Characteristically constant, diffuse, unresponsive to analgesia. **Altered sensorium** • Ranged from mild drowsiness to confusion and agitation, progressing to coma. **Seizures** • Frequently generalized tonic clonic may well start focally and often occur. Status epileptics has been reported. **Visual Symptoms** • Preceding visual loss and visual hallucinations entail occipital lobe origin. • Hemianopia, visual neglect and even Anton's syndrome. **Examination** • Fundoscopic exam often normal. • DTR may be brisk with Babinski signs present HTN is frequent but not always.
Investigations	**CT scan** • generally normal or may show less prominent non specific changes • symmetrical hypo-attenuation of white matter particularly in pareito-occipital distribution **Mri Brian with contrast** • More sensitive then CT brain in diagnosis • FLAIR sequence would show features of vasogenic oedema

Dr Nadeem Akhtar

	• Symmetrically in frontal parietal and occipital lobes. Temporal lobe is not commonly involved
	• Complete resolution within days to weeks after effective treatment
Pearl	• Distribution is not usually confined to a single vascular territory
	• DWI aids to distinguish PRES from a top of the basilar stroke
	• Calcarine and paramedian areas of occipital lobe are spared which helps to distinguish from bilateral occipital infracts

Management

General Measures	• High degree of suspicion to recognise the condition
	• Remove or control the underlying offending insult
	• ICU admission for monitoring and supportive care in severe cases
	• Seizures - Phenytoin, Can probably be safely tapered as sx and neuro-imaging findings resolve, usually after 1-2 wks.
Management of hypertensive PRES	• Blood pressure should be reduced cautiously
	• Use of IV Labetalol or Nicardipine to attain desired BP. If gbv neither agent effective, may consider nitroprusside, with caveat of theoretical concern of paradoxically increasing intracranial pressure through vasodilatation
	• Lower diastolic BP to 100-105 mmHg within 2-6 hrs. Max initial fall of BP should not exceed 25% of presenting value.

Non-hypertensive PRES

Cytotoxic Immunosuppressive Therapy	• If one of the chemotherapeutic agent is responsible for PRES then if possible it should be withdrawn quickly in consultation with oncologist
	• Delay may cause long-term damage to blood brain barrier.
	• Avoid the reintroduction of same agent if possible.
In setting of eclampsia:	• Delivery of baby and placenta are usually sufficient, but if not tx with Mg as opposed to Phenytoin or diazepam
	• No indication pts at long-term risk for sz recurrence or epilepsy (although long term studies are lacking)
Prognosis	• The anatomical extent of MRI findings have been shown to correlate with patient outcome.
	• Death may result from progressive cerebral edema, intracerebral hemorrhage or as complication of underlying condition.
	• More extensive brain involvement, particularly in brainstem correlates with worse prognosis

Reversible cerebral vasoconstriction syndrome

Overview/criteria

RCVS is characterised by recurrent thunderclap headaches with or without other recurrent neurological symptoms due to dysregulation of vascular tone resulting in constriction and dilatation. Large and medium size artery are mainly affected. CSF examination is typically normal.

- Female> man
- Ratio is about 3:!
- Mean age of onset is 42 years
- Reported ages include 10 -70 years

Risk factors

- Spontaneous (no identified exposure)
- Vasoactive drugs (more than half of the cases)
- Recreational drugs (cocaine, cannabis)
- SSRI
- Postpartum period
- Nasal decongestants
- Triptan
- Herbal Medicines.

Clinical Features	**Thunderclap headaches**
	▪ Sudden onset headache peaking within a minute.
	▪ Total of 1 -4 attacks in 1-3 weeks
	▪ Rarely may have only single attack.
	▪ Associated with nausea, vomiting, blurring of vision and sometimes with collapse.
	Transient focal deficits
	▪ Recurrent neurological deficit (10% of cases) lasting between few minutes to few hours (1 minute to 4 hours)
	▪ Can be visual, sensory or motor.
	▪ Transient dysphasia have been described
	Seizures uncommon
	✍ Neurological Equation Severe headaches (can be multiple thunderclap headaches + transient neurological deficits+ characteristic string and bead like appearance+ resolution of radioligcal findings in 1-3 months =RCVS
Investigations	**Non – contrast CT**
	▪ In uncomplicated RCVS: usually normal
	▪ May show cortical SAH or intracebral hemorrhage in complicated cases
	lumbar puncture
	if CTB is normal, to rule out SAH and inflammatory conditions like infection or cerebral vasculitis

MRI Brian

- Usually normal

- May show evidence of infarctions, especially in "watershed" zones May look like PRES

- Parenchymal hemorrhages or cortical SAH

MRA

- Alternating areas of constriction and dilatation – "beading" -- in several vascular territories

- May be seen in large-to-medium-sized arteries of anterior or posterior circulation

- Abnormalities may be absent early but show up on repeat imaging, believed to start distallyand move centripetally

- Not specific for RCVS

- Resolution within 3 months is most specific for RCVS

Management

General Measures

- Observation- Observation is adopted in some cases

- Symptomatic (pain, seizures, blood pressure control)

- Trigger avoidance (either activity or vasoactive substances)

Specific Measures

- Calcium channel blacker is used in many cases

- Nimodopine (30 -60 mg every 4 hour) tapered over few weeks.

- IV magnesium

Complications	▪ Early complications, (within the first week)
	▪ Localized cortical, non aneurismal, convexity SAH (20-25%)
	▪ Ischemic (mainly in arterial watershed regions) or hemorrhagic stroke (5-10%)
	▪ TIA (16%)
	▪ PRES
	▪ Permanent sequelae of a usually benign entity
	▪ Seizures
Pearl	▪ Overlap: about 10% of cases of RCVS are associated with PRES, regardless of cause. Share similar clinical features
Prognosis	▪ Highly dependent on the occurrence of stroke (6-9%)
	▪ Otherwise, by definition, most resolve completely without any sequelae

ADEM
Acute Disseminated Enecphalomylitis

Overview	Immune mediated Inflammatory and demyelinating condition of the central nervous system involving sub cortical white matter causing a mono-phasic illness and generally a favourable prognosis. characterised by typical MRI findings
Risk factors	• Any age can be affected • More common in children- 80 % of cases occur in less then 10 years old • Both sexes are affected • post infection -Occurs about 2 weeks (range 6 days to 6 weeks) • Post vaccination-up-to three months post vaccination • Post- Infection- in 50-70 % of cases it is preceded by viral or bacterial infection (immune attack is possibly triggered by molecular mimicry)
Clinical Features	**Monophasic** • ADEM can be defined as a generally a monophasic diseases. **Polysymptomatic Multifocal** • Polysymptomatic presentations could be variable combination of • Headache (Meningismus). • Optic Neurtis (often bilateral) • Acute transverse myelitis (frequently at thoracic levels). • Encephalopathy • Focal signs

	Meningoencephalitis like clinical picture.
	- Alteration in sensorium with
	- Fever
	- Headache and
	- Neck stiffness with other meningeal signs
	- This clinical spectrum is more common in children compared to adults
	Recurrent ADEM
	- Typically monophasic presentation, however some cases of recurrent ADEM has been reported
Investigations **Lumbar Puncture**	**CSF**
	- Essentially done to exclude any concomitant CNS infection
	- ADEM typically shows lymphocytic pleocytosis, (initially it can be polymorphonulear leucocytosis)
	- Elevated Protein (usually less then I mg/L)
	- Glucose (is usually normal)
	- Elevated IgG index
	- OCB are generally absent
	MRI Brain
	- Asymmetrical multifocal lesions in white matter on T2 and FALIR weighted sequence
	- Lesions are usually large > 1cm located in supratentorial and or infratentorial white matter regions
	- Involvement of grey matter including thalami and basal ganglia

	- Usually most of lesion enhance post contrast
	- No mass effect
	- Involvement in order of frequency
	- Sub-cortical white matter
	- Brian stem
	- Basal Ganglia
	- Cerebellum
	- Thalamus
Pearl	**Unlike MS :**
	- Relative periventricular sparing-is typical of ADEM
	- Basal ganglia and thalamus is frequently involved
	- Corpus callosum is usually not involved
	Spinal cord MRI
	- Large confluent intramedullary lesion(s) with variable degree of contrast enhancement
Management	
intravenous methylprednisolone	- IVMP- one gram for 3-5 days
	- Followed by oral predinsolone 1 mg/kg taper over 3 to 4 weeks.
	IV immunoglobulin
	- IVIG 2 gm/kg intravenously per day for 3-5 days
	- Can be a choice of treatment if infection (meningo-encephalitis) cannot be excluded hence steroids cannot be used.
Other options	**Plasmapheresis**

Response Monitoring	<ul><li>Recovery over 4-6 weeks 60-90% with no residual deficits.</li><li>Repeat MRI 6-12 months later to assess for lesion resolution</li></ul>
Pearl	NMO should also be considered in patients with prominent myelitis or ON, particularly if they are NMO-IgG positive

NDMAR ENCEPHALITIS

Overview	NMDA recptr encephalitis is an autoimmiune/paraneopalatic chanelapthy associated with antibodies to NR1/NR2 hetermoers of receptor commonly present as a neuropsychiatric symptoms in a young women
Risk factors	Women are commonly affected (80 %) More common in black women Median age 23 years Age range from 1 month to 65 years
	■ Testicular teratoma
	■ Medstinal teratoma
	■ Sex cord stromal tumor
	■ Small lung ca
	■ Hodgkin's Lymphoma
Clinical Features of Antibodies that affect Cell surface/ synaptic receptors	**Prodromal**
■ NMDAR	■ Agitation, psychosis.
■ AMPAR	■ Catatonia.
	■ Memory deficit, speech reduction.
■ GABA(B)	■ Abnormal movements.
■ LGI1	■ Seizures
■ Caspr2	**Clinical stage**
	■ Psychosis.
	■ Limbic encephalitis Dyskinesias.
	■ Seizures:
	■ Reduced level of consciousness.
	■ Sleep dysfunction

	Advanced stages.
	■ Coma.
	■ Hypoventilation.
	■ Dysautonomia.
Investigations	
Neurophysiology	**EEG-** ■ Generalised slowing in 70-90% of cases ■ 25-30% of cases would have specific changes of extreme delta brush.
Serology	**NMDAR Antibodies Serum and CSF:** ■ Higher Sensitivity in CSF ■ NMDA antibodies can be present only in CSF in about10-15 % ■ Sensitivity and specificity of 100 % ■ Antibody levels correlate with disease activity
Neuro-Imaging	■ MRI - can be normal in up to 25- 50 % of cases ■ T2 and Flair increased signals in cerebellar and medial temporal lobe ■ Faint and transient enactments can occur ■ DWI sequence- May show more widespread changes
General Imaging	**Pelvic Ultrasound** ■ Initial screen for ovarian teratoma **CT chest abdomen pelvis/PET** ■ To screen for other tumours

Management

General Measures	• Symptomatic treatment
	• Catatonia- IV Benzodiazepine (Lorazepam 2mg every 6 hours, maximum of 20-30 mg in 24 hours)
	• ECT- in refractory and severe cases of catatonia cases
Specific Measures	Early recognition is of paramount importance
	• Definitive treatment is combined approach to treat tumour (resection) along with immuno-supression
	Immuno-suppression
	• Steroids
	• IVIG- 0.4 mg/kg for 5 days
	• Plasma exchange
	Second line of treatment
	• If there are no improvement after 7-10 days then consider immune-suppressive treatment includes
	• Rituximab 375 mg/m2 every 4 weeks
	• Azathioprine
	• Mycophenolate
	• Cyclophoshomide 750 mg/m2 every one month
	• In autoimmune NMDA receptor encephalitis Imunno-suppresion should continue for minimum of a year
	Tumour surveillance
	• Tumour Surveillance – every 6 months for 3 to 4years

Rehabilitation	▪ Recovery is a slow
	▪ Comprehensive multidisciplinary rehabilitative
	▪ Speech therapist
	▪ Occupational therapy
	▪ Physiotherapy
Pearls	▪ Antibodies associated to epilepsy (usually non–paraneoplastic) Synaptic intracellular Amphiphysin; GAD65
	▪ Limbic encephalitis, Refractory seizures
	Other antibodies and their clinical phenotype Intracellular
	HuD Ma2 CRMPS
	◆ Encephalomyelitis
	◆ Limbic encephalitis
	Caspr2-
	◆ Encephalitis and/or neuromyotonia, Morvan's syndrome, painful neuropathy
	GlyR
	◆ Stiff-perosm syndrome, hyperekplexia
	PERM mGluR
	◆ Ophelia syndrome
	DPPX (Kv4.2)
	◆ Agitation, myoclonus. tremor, seizures, diarrhoea
Prognosis	▪ Good recovery in 80 % of cases
	▪ Relapse occurs in about 10 % of cases
	▪ Mortality rate- 5-7% of cases

Dr Nadeem Akhtar

Cauda-equina Syndrome

Overview

Cauda-equina Syndrome

Peripheral nerve injury to lumbar, sacral and coccygeal nerve roots Symptoms:

Variable motor and sensory loss in lower extremities sciatica bowel and bladder dysfunction. Saddle anasthesia

Risk factors

Acute causes

- Central disc herniation
- Vertebral collapse due to metastatic infiltration
- Spinal subarachnoid Haemorrhage
- Acute extra Dural haematoma
- Trauma

Chronic causes

- Primary tumors arsing form spinal (ependydoma and neurfibroma)
- Extrinsic tumors
- Achndroplasia
- Degenerative spondylosis

Clinical Features

- Mild back discomfort
- All patients who complain of urinary or fecal incontinence should be considered to have CES until proven otherwise
- pressure sensation" in lower back
- Numbness in perineal area
- Unable to void urine
- Fecal incontinence

Signs	**Physical** - normal power examination of lower - extermities - Diminished peri-anal sensation - Sphincter tone is decreased in 60 to 80% of patients - Inability to actively contract anus - Urinary retention / overflow incontinence
Investigations	**Electrophysiological testing.** - Standard nerve conduction and EMG studies of lower limbs to determine the damage to nerve roots within cauda equina
Pearl	- Evaluation of the urinary post-void residual volume assists with diagnosis: the absence of a post-void residual volume of over 100ml, essentially excludes a diagnosis of CES, with a negative predictive value of 99.99%
Management	Outcomes related to permanent damage to nerve roots This damage occurs within minutes to few Hours - Neurosurgical consultation - High dose systemic corticosteroids - Emergent surgical decompression Early vs. late decompression No clear evidence that there is a significant difference in outcomes.

Dr Nadeem Akhtar

Status epilepticus

Overview	**Status epilepticus** Traditionally defined as continuous, unremitting seizure lasting longer than 30 minutes,or recurrent seizures without regaining consciousness between seizures for greater than 30 minutes Newer literature and common neurology practice usually accepts 5 minutes or longer as definition for status epilepticus.
Risk factors	Status epileptics • Idiopathic: no acute precipitating CNS or metabolic insult- • Acute symptomatic: Due to an acute illness with known CNS insult or metabolic disturbance • Remote symptomatic: no obvious acute precipitant, but prior known CNS insult is well known to increase seizure risk.
Clinical Features Signs	• Primary or secondary generlaised • Symmetrical gerlaised tonic and clonic activity • Imaparietn of the conciousnesss • Varaible post ictal period
Investigations	• FBC • Eletrolytes • Check serum drug levels: Phenytoin, valproate, carbamzepine ect • Blood and urie for toxicology screen. If no obvious cause is found

Management

General Measures

- **ABCDE**
  - Monitor and mantian airway- if GCS is redueced and at risk of aspiration
  - Oxygen therapy- be aware that patient may require intubation.
  - Immediate IV access
  - Imeediate Glucose levels
  - Check electrolytes- Na ,Ca Mg PO4
- Start IV infusion with saline solution.
- Administer 100 mg thiamine, IV.
- Administer 50 ml of 50% glucose solution, IV, if blood sugar is low or unobtainable.
- Do not give glucose if blood sugar is normal or high.
- Always have CPR equipment at bedside of a patient in status.
- Diazepam- 5mg
- Lorazepam-2mg
- Further repetation-every 4 miutes rate of termination.Lorazepam 59.1%Diazepam 42.6%

Diazepam

- Rapid crossing of Blood brian barrier due to highl lipid solubility helps in controlling in 1-3 minutes
- Rapid fall of concentration in brian
- Rapid distribution in fatty tissues
- Diazepam Load 0.2mg/kg (20mg rectally or 10mg IV)

Lorazepam
- Lipd slubiltiy is less than diazepam

  - Slower passage through blood brian barrier, generally stops seizures within 6-10 min

- Not as rapidly redistributed to fat stores

  - Longer duration of action 12-24 hr
  - Half life T1/2 =14 hr
  - Dose: 0.05–0.1mg/kg

Pearls
- Lorazepam is benzodiazepine of choice because of smaller volume of distribution, Longer therapeutic half life, Anti-convulasnt effect lasts for 6-12 hours, compartively less respiratory depression than diazepam

Midazolam

  - Can be given intra-muscular
  - Midazolam Load IM 0.2mg/kg up to 10mg IM

Initial anti-epileptic drug treatment:

Phenytoin	
	<ul><li>Cardiac monitoring necessary</li><li>IV dosing 20 mg/kg load</li><li>A dose of 15mg/kg at a rate of 50mg/min in young, 20-30mg/min in elderly.</li><li>If already on phenytoin, give 10mg/kg, and request urgent levels.</li><li>A further 5mg/kg given almost immediately afterwards if no response to initial dose.</li><li>Can give upto 30mg/kg., before consideration anaesthesia.</li></ul>
	<ul><li>Max infusion rate of 1mg/kg/min, which is max- 50 mg/min</li></ul>
	<ul><li>15-20minutes at least to take effect</li><li>Stops seizures in 10-30 minutes</li><li>Must be aggressive with dosing, and even if responds to initial dose, should aim for high normal levels. Range 40-80 micromoles/litre.</li><li>Request serum phenytoin level 1/2hr to 1 hr after loading dose</li><li>In general for micromoles/l, for every 4 micromoles/l off desired target, give 1mg/kg</li><li>Duration of action is generally 24 hrs,</li><li>T ½=24hr</li><li>Side Effects: arrhythmias, hypotension, wide QT interval, phelibitis</li></ul>

Pearl	▪ Common problem is that about 70% patients admitted to ITU are given an inadequate loading dose.

Fosphenytoin- phenytoin prodrug

- ◆ IV dosing: 20 mg/kg load
- ◆ Safer than phenytoin
- ◆ May give IV or IM
- ◆ May give faster than phenytoin (100-150mg/min)
- ◆ Much more expensive

Sodium Valproate

- ◆ Has been reported to be effective in GTCSE
- ◆ Rapid loading dose appears safe
- ◆ 25-30mg/kg rapidly infused
- ◆ Side Effects: dizziness, HA, nausea
- ◆ Efficacy rates 63% in one study.
- ◆ Loading dose 25-45mg/kg.
- ◆ Rate 200-500mg/min.
- ◆ Continuous infusion rate upto 6mg/min.

Consider Keppra IV Load

- — Off label
- — Levetiracetam 1500-4000mg infuse 500mg/min
- — use 30mg /kg IV
  - ◆ Lacosamide 400mg IV infuse over 5 mins

	Oral Agents/Via NG • Topiramate 400-800mg/day (metabolic acidosis • Pregabalin 150-600mg per day • Vigabatrin 0.5-1gr q8h (no short term side effects)
Refractory Status	▪ intubation, IV access ▪ Continuous EEG monitoring ▪ If seizures persist, the patient should be placed in a drug induced coma with pentobarbital, a benzodiazepine, or other anesthetic agent to prevent life threatening lactic acidosis, hypoxia, hyperthermia, and permanent seizure-induced neuronal damage. ▪ The patient must be in an ICU, and outcome should be monitored and treatment guided by EEG with the goal being suppression of seizure activity on EEG.
	▪ Anaesthetic agents • Midazolam • Propofol • Pentobarbital
Pearl	▪ Repeated failed attempts at withdrawal anaesthesia. One would test (Even with known epilepsy) should have MRI and CSF as a minimum. ▪ Serum amticonvulsant levels. ▪ Alternative anti-epileptics • iv valproate worth a go • Leviteracetam • Topiramate ▪ Anoxic / metabolic brain damage.. Post-anoxic myoclonus?

Dr Nadeem Akhtar

Prognosis	• Mortality and morbidity severely influenced by underlying aetiology. Cannot give reliable figures for condition itself. Mortality 20%
	• Morbidity; high risk recurrent seizures, cognitive deficits, and future episodes

Acute transverse myelitis (ATM),

Overview	Acute transverse myelitis (ATM), an inflammatory myelitis, is one of the causes of acute transverse myelopathy.
	• Focal inflammatory disease process of the spinal cord. Affects motor, sensory and autonomic fibers
	• Highest incidence between 10-19 yo, 30-39 years (bimodal)
	• The differential diagnosis of acute inflammatory transverse myelitis (ATM) is broad. Therefore, physicians must be aware of the many potential etiologies for acute myelopathy.
Risk factors	• Up to 50% with preceding febrile illness
Clinical Features	• The differential diagnosis of acute inflammatory transverse myelitis (ATM) is broad. Therefore, physicians must be aware of the many potential etiologies for acute myelopathy,
	Complete transverse myelitis
	• Present with usually has more or less symmetrical findings and involvement of motor, sensory, and sphincter function
	• Acute/subacute onset of pain, weakness, sensory loss,
	• Incontinence over hours-weeks
	• 80% of cases are in thoracic spine
Signs	Neuro exam:
	• Depends on where the lesion is!
	• Typically bilateral weakness with UMN signs, but

	▪ If anterior horn cells are involved, may get a predominately LMN exam Incomplete transverse myelitis ▪ Asymmetric findings that may involve a limited number of tracts and does not typically result in loss of all motor, sensory and sphincter function.
Investigations	
Serology	**Autoimmune-NMO-IgG** ▪ Thirty-eight percent of patients with a first episode of transverse myelitics were seropositive for NMO-IgG
Lumnar Punctur	**CSf** ▪ Pleocytosis, elevated IgG index MRI Spine, with and without contrast Labs: to rule out other disease processes
	▪ Serologic of connective tissue disease ◆ Sarcoidosis, ◆ Behcet's disease, ◆ Sjogren's syndrome, ◆ SLE, ◆ mixed connective tissue disorder ▪ Laboratory evidence for ◆ Syphilis, ◆ Lyme disease ◆ HIV,

Dr Nadeem Akhtar

	• HTLV-1,
	• Mycoplasma,
	• Other viral infection (e.g. HSV- 1, HSV- 2, VZV, EBV, CMV, HHV- 6, enterovirus)
Management	▪ Treatment: IV steroids + PT
	▪ Outcome:1/3 no long term sequelae, 1/3 mild-moderate disability, 1/3 severe disability
Pearls	▪ Inflammation within the spinal cord demonstrated
	by CSF pleacytosis or elevated IgG index or gadolinium enhancement if none of the inflammatory criteria is met at symptom onset, repeat MRI and lumbar puncture evaluation between 2 and 7 days following symptom onset
	▪ In MS, lesions are usually small (< vertebral segments in length) and peripheral, and therefore cause asymmetric symptoms and signs.
	▪ Patients with Complete Transverse Myelitis in general, are at low risk for future development of MS. However, they could have recurrences consistent with relapsing myelitis or NMO.
	▪ Two autoimmune markets that may predict recurrence are anti-Sjo"green's syndrome antibody (SS-A) and NMO-IgG
Prognosis	▪ 1/3 recover without any sequlae
	▪ 1/3 have moderate impairment
	▪ 1/3 have permanent disability

NMO (Neuromyelitis optica)

Overview	▪ The traditional concept of neuromyelitis optica was that it is a monophasic disorder, in which near-simultaneous bilateral optic neuritis and transverse myelitis arise. With MRI evidence of myelitis involving 3 or more contiguous segments OR NMO-IgG + aquaporin 4, found in serum
	▪ Neuromyelitis opticaq is recognized as a discrete, relapsing, demyelinating disease, with clinical, neuroimaging, and laboratory
Risk factors	▪ Several environmental and genetic factors
Clinical Features	▪ Neuromyelitis optica is recognized as a discrete, relapsing, demyelinating disease, with clinical, neuroimaging, and laboratory
Signs	▪ **Optic neuritis** ◆ Acute/subacute neuropathic visual loss ◆ Typically painful ◆ Mild, if any, disc edema ▪ **Myelitis** ◆ Acute/subacute weakness, numbness ◆ Bowel/bladder problems ◆ L'hermitte's Sign ◆ Monophastic,>70% recurrent ON and SC attacks may be in years apart

	• Limited forms of neuromyelitis optica
	◆ Optic neuritis: recurrent or simultaneous bilateral
	◆ Asian optic-spinal multiple sclerosis
	◆ Optic neuritis or longitudinally extensive myelitis associated with systemic autoimmune disease
	• Asian optic-spinal multiple sclerosis
	◆ Optic neuritis or longitudinally extensive myelitis associated with systemic autoimmune disease
Pearl	• Findings that can distinguish it from multiple sclerosis, simultaneous ON and Ssease Bilateral ON involvement
	• Aquaporin-4, IgG autoantibody localizes to glia at blood-brain-barrier Binds to aquaporin-4, the main water channel in the central nervous system
Investigations	Aquaporin-4,
Serology	IgG autoantibody localizes to glia at blood-brain-barrier Binds to aquaporin-4, the main water channel in the central nervous system
	• About 90% specific, 75% sensitive for NMO
	• Often + in brain MRI-negative relapsing myelitis/optic neuritis

Dr Nadeem Akhtar

Lumbar Picnture	**CSF**
	<ul><li>>50 white blood cells/mm3 or >5 polymorphonuclear Leucocytes/mm3</li><li>Oligoclonal bands</li><li>IgG synthesis less common</li></ul>
Neuro-Imaging	**MRI Spinal Cord**
	<ul><li>Elongated, expansile, enhancing spinal cord lesions</li><li>Idiopathic single or recurrent events of longitudinally extensive myelitis(>3 vertebral segment spinal cord lesioni seen on MRI)</li></ul>
Brain MRI	**Brain MRI**
	<ul><li>Usually normal; occasional multiple-sclerosis-like plaques or confluent/symmetrical lesions</li><li>Optic neuritis or myelitis associated with brain lesions typical of neuromyelitis optica(hypothalamic, corpus callosal, periventricular, or brainstem.</li><li>Can be detected with MRI occur in 60% of patients with neuromyelitis optica</li><li>Later in the course of the disease but these lesions are usually clinically silent.</li></ul>
Management	
General Measures	Intravenous corticosteroid therapy
	<ul><li>is commonly the initial treatment for acute attacks of optic neuritis or myelitis.</li><li>IV methylprednisolone 1000 mg/day, 3-5 days</li></ul>

Specific Measures	Plasmapheresis
	<ul><li>Patients who did not respond promptly to corticosteroid treatment benifited from seven treatments of plasmapheresis (1-0 to 1-5 plasma volume per exchange over a period of 2 weeks</li><li>Early initiation of plasmapheresis is recommended, particularly for patients with neuromyelitis optica with severe cervical myelitis, who are at high risk for neurogenic respiratory failure.</li><li>Plasmapheresis is also beneficial for patients with acute, severe vision loss who have optic neuritis that is refractory to corticosteroid therapy.</li></ul>
	Prevention of Recurrence and atabiliza tion
	<ul><li>Azathioprine 2.5-3 mg/Kg/day</li><li>Concurrent prednisone 1 mg/Kg/day, tapering slowly after azathioprine takes effect</li><li>Mycophenolatemofetil,Mitoxantrone, Rituximab, IVIg, Plasmapheresis possible second liners</li></ul>
Pearls	<ul><li>Patients with incomplete Transverse Myelitis</li><li>This group of patients is currently regarded as having a clinically isolated syndrome (CIS), which places them at risk for developing other symptoms that will lead to a definate diagnosis of MS.</li><li>Older patients with more common, sequential optic neuritis/myelopathic disease</li></ul>

Prognosis	• The prognosis for MS attacks may be much better than for NMO attacks, and
	• Less severe disease at onset
	• High-titer + NMO-IgG antibodies
	• Step-wise progression portends worse prognosis than monophasic disease

Limbic encephalitis

Overview	Can be divided into two types
	1. Autoimmune Limibic Encephalitis-VGKC associated
	2. Paraneoplastic limbic encephalitis (PLE)
	VGKC complex antibodies
	▪ GABA-B—Limbic encephalitis with prominent seizures, status
	▪ AMPA—Limbic encephalitis, psychosis LGI1—Limbic encephalitis, myocionus, hyponatremia,
	▪ Limbic encephalitis, hyponatremia, myoclonic-like movements (tonic seizures, facia-branchial dystonic seizures
	Paraneoplastic linbic encephalitis (PLE)-
	▪ Anti Hu, Anti Ma
Risk factors	▪ Several envormentlal and genetic factors
Clinical Features	Typically presents with subacute development of
	▪ Subacute memory impairment,
	▪ Confusion,
	▪ alteration of consciousness,
	▪ Disorientation, and behavioral change attributable to limbic dysfunction often accompanied by seizures and temporal Lobe origin

	- An appropriate clinical phenotype developing over a maximum of 12 weeks (although proven cases with longer courses have been described)
	- Neuropsychological testing, where possible, reveals fronto-temporal dysfunction with prominent episodic memory impairment and relative sparing of parietal lobe function. - Hyponatraemia due to the syndrome of inappropriate antidiuretic hormone secretion (SIADH) VGKC associated limbic encephalitis antibodies - There is much evidence to suggest that in VGKC associated limbic encephalitis the antibodies are pathogenic - VGKC antibodies are found in CSF and serum, and there is a temporal relationship between treatment, reduction in antibody levels, and clinical improvement.
Investigations	
Lumbar Puncture	CSF
Neuro-Imaging	MRI Brain - Not all patients have temporal lobe MRI signal change; there is some evidence to suggest that
Gneral Imaging	CT chest abdomen and Pelvis (to look for underlying malignancy) PET (Paraneoplastic limbic encephalitis) - PET imaging may be useful in demonstrating temporal lobe abnormalities in MRI negative cases - Pick up tumour elsewhere

Pearl	▪ Paraneoplastic limbic encephalitis may present months or even years before the detection of a tumour, and may occur in the absence of paraneoplastic antibodies. Whole body FDG-PET may be superior to CT imaging alone to detect occult neoplasia

EEG

Management

General Measures	▪ Infections must always be considered first and empirical treatment should not be delayed if there is any doubt about the diagnosis. ▪ Although a wide range of viral, bacterial, and tropical infections, and even neurosyphilis
Specific Measures	Paraneoplastic LE- ▪ Mainstay of treatment is identification and elimination of tumour. ▪ If no tumour is identified then suspicion for a possible underlying cancer should remain and earlyrepeat cancer screeningshould be performed. ▪ Once tumour is identified , Refer to oncologist for tumour specific treatment ▪ Immunmodualtion(IVIG/PLEX/ Steroids) is generally not helpful ▪ But can be considered along with or in conjunction of ontological treatment ▪ Few case reports suggestrituximab was found to be helpful in some cases. Of Paraneoplastic LE.

Pearl	▪ PLE and related syndrome precede identification of tumors which are generally in their early stages. hence they provided wind of opportunity to be identified and treated successfully Autoimmune LE- ▪ Hyponatermai (Associated mainly with LG1) ▪ Plasma exchange or IVIG is considered to be first line ▪ Alternatives - Pulsed Methyl predisnolone= for six months ▪ Maintenance Steroids - 1mg/kg for few weeks slow wean off over ▪ If relapses then- consider use of Azthioprine for
Pearls	▪ Patients with herpes simplex encephalitis typically present with a fairly abrupt onset of confusion, memory impairment, and often seizures. ▪ Fever is common but not invariable. ▪ Herpes simplex is not only the commonest identified cause of viral encephalitis in general, but by far the commonest cause of viral limbic encephalitis in particular paraneoplastic limbic encephalitis.

Dr Nadeem Akhtar

Gullian Barre Syndrome

Overview	• The typical illness evolves over weeks usually following an infectious disease and involves. Fairly symmetric weakness in the legs, later the arms and, often, respiratory and facial muscles
Risk factors	GBS Antecedent infections • About two-thirds of patients have symptoms of an infection in the 3 weeks before the onset of weakness. Vaccination and other events • Many reports have documented the occurrence of GBS shortly after vaccinations, operations, or stressful events, but the specific relation with GBS is still debated
Clinical Features	Flu like symptoms or diarrhoea • 1-3 weeks prior to neurlogical symptoms can occur in almost 2/3 of cases
Signs	Paralysis • Fairly symmetric ascending paralysis. with typical involvement of proximal muscle groups and then distal muscles • Pace and extent of progression differ, but in severe cases there is marked quadriparesis in addition to bilateral facial weakness • 10% will first develop weakness in face or arms - severe resp muscle weakness in 10-30% pts -oropharyngeal weakness in 50%`

	Pain
	- Often prominent severe pain in lower back, Common to have paresthesias in hands and feet
	Dysautonomia
	- is common: tachycardia, urinary retention, hypertenison alternating w/ hypotension, ileus may lead to a grave outcome
Investigations	Antiganglioside antibodies
Serology	- Acute infi ammatory demyelinating polyradiculoneuropathy (AIDP) – Unknown
	- Acute motor (and sensory) axonal neuropathy (AMAN or AMSAN) – GM1, GM1b, GD1a, GalNac-GD1a
	- MFS and GBS overlapping- GD3, GT1a, GQ1b, in about half of patients with GBS, Serum antibodies to varous gangliosides have be found in human peripheral nerves, including LM1, GM1, GM1b, GM2, GD1a, GalNAc-GD1a, GD1a, GD2, GD3, GT1a, and GQ1b.
Lumbar Puncture	CSF
	- Album inocytologic dissociation: elevated CSF protein w/ normal WBC (80-90% pts)
Neurophysiology	Electromyography (EMG) helps to confirm diagnosis-
	• Prolonged or absent F waves (indicating involvement of proximal parts of nerves and roots) and reflecting focal demyelination,
	• Reduction in the amplitude of muscle action potentials.
	• Slow nerve conduction velocity

	▪ Conductino block in motor nerves
	▪ Prolonged distal latencies (reflecting distal conduction block)
Management	General Measures
	▪ Requires regular assessment of vital signs every 2 -4 hours or intensive nursing care
	▪ Swallowing assessment
	▪ Cardiac monitoring in all patients who are severely affected
	▪ DVT prophylaxis
	▪ Neuropathic pain, pain control with Gabapentin or carbamazepine may be used during the
	▪ Monitor Respiratory status closely, up to 30% may required ventilator support
	▪ BP Monitoring
	▪ In severe cases, intrarterial monitoring may be necessary given the significant blood pressure fluctuations
	▪ HTN, can be treated with Labetalol
	Indications to start IVIg or PE:
	▪ Severely affected patients (inability to walk unaided)
	▪ Start IVIg preferably within first 2 weeks from onset: 0.4 g/Kg for 5 days or 5* PE with total exchange volume of five
	▪ Plasma volumes in 2 weeks. Plasma exchange, 1 volume q 1-2 days for five exchanges;use albumin as replacement fluid unless coagulopathy develops

	Indications for re-treatment with IVIg;
	• Secondary determination after initial improvement or stabilisation (treatment-related fluctuation): treat with 0.4 g/Kg for 5 days
	• No proven effect of re-treatment with IVIg in patients who continue to worsen
Pearl	• The choice b/w plasma exchange and IVIG is depends on availability, pt contraindications, etc. Because of ease of administration, IVIG is frequently preferred. The cost and efficacy of the 2 treatments are comparable.Glucocorticoids have NO ROLE!!
ITU admission if	• FVC <40 mL/Kg;
	• Poor cough (predict respiratory muscle weakness)
	• Bulbar dysfunction (Dysphagia and inability to protect airway)
	• Failed swallow evaluation (increased risk of aspiration)
	• Automatic instability/Blood Pressure lability
Prognosis	• 65% can walk independently @ 6 months
	• Overall, 80% usually recover completely
	• 5-10% have prolonged course W / incomplete recovery, 3% wheelchair bound
	• Approx 5% die despite ICU care
	• 2% will develop chronic relapsing Chronic Inflammatory Demyelinating Polyradiculoneuropathy (CIDP)

Dr Nadeem Akhtar

Pearls	Review the diagnosis of GBS if
	<ul><li>Severe pulmonary dysfunction with limited limb weakness at onset</li><li>severe sensory signs with limited weakness at onset</li><li>Bladder or bowel dysfunction at onset</li><li>Fever at onset</li><li>Sharp sensory level</li><li>Slow progression with limited weakness without respiratory</li><li>Involvement (consider subacute infl ammatory demyelinating polyneuropathy or CIDP)</li><li>Increased number of mononuclear cells in CSF (>50*106/L) polymorphonuclear cells in CSF</li></ul>

Miller Fisher Syndrome

Overview	Triad of acute external ophthalmoplegia+ ataxia, + areflexia (minimal or no motor or sensory deficit)
	▪ Miller Fisher Syndrome (MFS): is the most common GBS variant
	▪ Currently MFS varient of GBS is thought to occur in up to 5% of all GBS cases in Western countries, with higher incidence in Asian countries
	▪ Incidence of roughly 1/1,000 000 The onset of MFS varies from 13 to 78 years of age with a mean of 43.6 years
Risk factors	▪ Recent flu like illness
	▪ Antecedent infections
	▪ Vaccination and other events
Clinical Features	Rapidly evolving triad of
	▪ ataxia,
	▪ areflexia,
	▪ ophthalmoplegia with or without mild limb weakness
	Diplopia
	▪ The common initial symptom is diplopia due to bilateral extraocular muscles weakness with horizontal, followed by vertical, gaze inability.
	▪ Eyelid ptosis may be present at the peak of disease course in more than half of cases but pupillary function is usually spared.
	Ataxia
	▪ Ataxic gait is a predominant symptom of MFS.

	Weakness • Muscle strength is usually preserved but superimposed incoordination from sensory ataxia may emerge.
Investigations	
Serology	GQ1b antibodies • Polyclonal anti-ganglioside Ab of IgM, IgA and IgG classes, found in serum of patient with acute MFS; • sensitivity and specificity > 90% (IgG is measured for clinical diagnostic purposes) • May be present in other GBS variants, such as • GBS with ophthalmoparesis, • Bickerstaff encephalitis, • pharyngo-cervical brachial GBS) • Ab levels peak at presentation, rapid decay with clinical recovery • Follows a variety of infections, including Campylobacter jejuni. • Molecular mimicry between GQ1b and C. jejuni lipooligosaccharide seems to be central to pathogenesis (clinical pattern of restricted muscle involvement likely due to patterns of ganglioside expression within specific muscle groups)
Electrophysiology	• Early-F waves are normal eaeliers in the course of illness; then it will get prlonged • Later- Prolonged distal latencies and increased duration/polyphasia of distal compound muscle action potential

Management

Specific Measures	No control trials to study effect of established treatments for GBS i.e plasma exchange or IVIG ▪ IVIG, 0.4 mg/kg once a day for five days ▪ Plasma exchange, one voule every 1-2 days for five exchanges
Learning points	▪ Wernicke encephalopathy also presents with ataxia and ophthalmoplegia but typically affects the lateral recti causing esotropia, and it is often seen in individuals with malnutrition, such as alcoholics with thiamine deficiency. ▪ Wernicke encephalopathy patients usually have apparent symptoms and signs of peripheral neuropath

Prognosis	▪ Gradual and complete recovery over weeks to months is common in most of MFS patients
	
	  ♦ Some MFS patients may have swallowing or respiratory problems or develop arm or leg weakness or autonomic instability and in this setting IVIG may be beneficial.
	
	  ♦ Time to recovery with often complete symptom resolution is in the region of 8-12 weeks.

Bickerstaff's brainstem encephalitis (BBE)

Overview	Bickerstaff's brainstem encephalitis (BBE): ophthalmoplegia, ataxia, altered consciousness, hyperreflexia Brain MRI shows brainstem hyperintensities
Risk factors	• Recent flu like illness • Antecedent infections • Vaccination and other events
Clinical Features Signs	BSE According to the BBE diagnostic criteria, all the patients had • external ophthalmoplegis and • ataxia (both truncal and limb) • During the illness, Consciousness was disturbed in 74% (drowsiness, 45%; stupor, semicoma or coma, 29%). • Limb weakness was appreciable in 60% • Deep tendon reflexes were absent or decreased in 58% normal in 8%, and brisk in 35% • Babinski's sign was present in 40%. • Bulbary palsy (34%)
Investigations Serology	• From an immunological perspective, anti-GQ1b IgG antibody is frequently detected in sera from Patients with
Lumbar Puncture	CSF • Albuminocytological dissociation

	▪ CSF pleocytosis was present in about one-third during
Neuro-imaging	MRI
	▪ Abnormal lesions (high-intensity areas on T2-weighted images of the brainstream, thalamus, cerebellum and cerebrum
	▪ Normal MRIs have been reported
Neurophysiology	Electrodiagnostic Criteria
	▪ Motor nerve dysfunction is predominantly axanol and the form of GBS that overlaps with BBE may be the axanol subtype, i.e. acute motor axanol neuropathy (AMAN)active denervation potentials
	▪ Invariably have low distal CMAPs, which could be caused by wither axonal degeneration or distal conduction block
	▪ There are no randomised controlled trials of immunomodulatory therapy in Fisher Syndrome or related disorders on which to base practice.
	▪ Intravenous immunoglobulin (IVIg) and plasma exchange are often used as treatments in this
	Emergency EEG
	▪ May be necessary to make diagnosis.
	▪ Look for diffuse non specific abnormalities or periodic lateralizing epileptiform discharges (PLEDS Characteristic EEG – Spike and slow-wave activity and
	▪ PLEDS(period lateralized epileptiform discharges), which arise from temporal lobe;

Management	• There are no reliable trials of immunomodulatory therapy in MFS/ BSE
	• IVIG and Plasma exchange are often use as first line treatment
Pearls	Clinical features of BBE with overlapping GBS
	• In the 37 patients who had BBE with limb weakness, muscle weakness was symmetrical and accid. The clinical diagnosis was therefore BBE with overlapping GBS. Twenty-ve had GBS with limb weakness of 4 on the MRC scale, and the other 12 had scores of <3.

Viral Encephalitis-Herpes Simplex

Overview	<ul><li>The most commonly identified cause of acute, sporadic viral encephalitis; 10 to 20 % of all case</li><li>Subtype 1 virus causes more than 95% of cases of HSV encephalitis</li><li>Children and young adults, primary HSV infection may result in encephalitis; virus enters the central nervous system (CNS) by neurotropic spread from the periphery via the oldfactory bulb</li></ul>
Risk factors	<ul><li>Age- 20-40</li><li>Imuncompromised</li></ul>
Clinical Features	<ul><li>Change in personality, altered mental status,</li><li>Decreasing level of consciousness</li><li>Fever</li><li>Headache</li><li>Seizures (focal and generalized)</li><li>Focal neurologic findings</li><li>Dysphasia, cranial nerve paresis,</li><li>hemiparesis</li></ul>
Investigations	
Neuro-imaging	CT head<ul><li>First 5 days: CT sensitivity 73%, specificity 89%</li><li>>5 days: CT sensitivity 90%, specificity 92%</li><li>Low density lesions of the temporal lobe:</li><li>(88%) in HSE</li></ul>

	• Edema in the temporal lobe • Hemorrhagic necrosis can cause midline shift MRI • Hyperintensity on T2 in one or both temporal lobes, which may extend into cortex
Lumbar Puncture	CSF • Opening pressure can be elevated in approximately one-third of patients • WBC range – 10 -= 100s, /mm3 75-100% lymphocytes • RBC <10 protein, 0.6-6 g/L Glucose moderately reduced in a small percentage
Neurophysiology	Emergency EEG • May be necessary to make diagnosis • Look for diffuse nonspecific abnormalities or periodic lateralizing epileptiform discharges(Pleds characteristic EEG – Spike and slow-wave activity and • PLEDS (periodic lateralized epileptiform discharges), which arise from temporal lobe; attenuation of background EEH:spike and slow wave activity from the temporal lobe. Sensitivity 85%. Specificity 33%.
Management	
General Measures **Specific Measures**	• O2, fluids, NG feed • Acyclovir(10mg/kgq8hr)immediately when diagnosis is suspected. • Continue for 14 days. It should only be discontinued if an alternative definite diagnosis is made.

Dr Nadeem Akhtar

	- If PCR negative and no other support for HSE, stop acyclovir
	- Renal toxicity may occur and needs to be monitored
	In immunosuppressed patients
	- Treat for 21days.
	Steroids
	- Can be considered if there is evidence of raised ICP
Prognosis	- Five percent of patients may relapse
	- Mortality > 70% if untreated (20% with Rx)
	- 2/3 rds pts have neuropsychiatric sequelae
	◆ 69% memory impairment
	◆ 45% personality/behaviour change
	◆ 41% dysphasia
	◆ 25% epilepsy
	- Poor prognostic factors
	◆ Age > 60 yrs
	◆ GCS < 7
	◆ Delay in starting aciclovir (esp > 2 days)

Subarachnoid Haemorrhage

Overview	▪ They occur in young people
	▪ 80% in 40-65 year olds
	▪ 15% in 20-40 year olds
	▪ It can kill quickly
	▪ 25% die within 24 hours
	▪ 50% will be dead at 6 months
	▪ It causes significant disability
	▪ Cognitive impairment
	▪ Neurological disability depending on size of bleed & complications encountered
Risk factors	▪ Higher chance if:
	▪ Female
	▪ 3rd trimester of pregnancy
	▪ Middle-aged
	▪ Abuse of stimulant drugs
	▪ Connective tissue disorder
	▪ Family history
	▪ PCKD
Clinical Features	Headache with altered or loss of consciousness
	◆ Worst ever headache of life
	◆ Thunderclap headache
	◆ With or without variable period of loss of concioussness
	◆ 50 % do not awaken

	On exmiantion
	♦ Neck stiffness
	♦ Photophobia
	♦ Subhylod heamrrohage-on fundsocpy
	Local pressure effects of the aneurysm
	▪ ACOM
	♦ Visual symptoms due to optic chiasm compression
	♦ Positive babinski
	♦ Bilateral lower limb paresis
	▪ MCA
	♦ Contralateral hand & face paresis
	♦ Contralateral visual neglect
	♦ Aphasia (dominant side)
	▪ ICA/Pcom
	♦ CNIII signs
Signs	▪ Normal exam
	▪ Confusion/memory loss
	▪ Aphasia
	▪ CN abnormalites
	— CNII – papilloedema, usually mild initially & retinal haemorrhages
	— CNIII – palsy
	— Hemiparesis/neglect
	Obs
	♦ HTN, tachycardia, febrile

Dr Nadeem Akhtar

Pearl	- Painful 3rd nerve palsy - Compression of the 3rd nerve by the PCA - The pupil is dilated – different from diabetes which typically spares the pupil

Investigations

Neuro-imaging	CT scan - Initial study of choice is an urgent CT scan withoutcontrast - Sensitivity decreases with time form onset of thunderclap headache - CT scan is 90 % sensitive within the first 24 hours, 80 % snesitve at 3 days and 50 % sensitive at one week CT angiogram - To rule out aneurysm
Lumbar Puncture	Lumbar Puncture - if the history is strongly suggestive and the CT is negative Lumbar puncture is performed - Xnathochromia is classic sign but not present elary 12 hours sensitivty

Management

Specific Measures	- Main aim is damage control – want to prevent further bleeding & try to avoid the complications that SAH patients get Identifying and treating the causative lesion, thus preventing re-bleeding Treating and preventing vasospasm - CalciumChannelBlockersNimodipine 60mg q6h x 24d • Reduces: – Neurologic deficit Cerebral infarction – Mortality

	- Blood vessel goes into spasm causing ischaemia - stroke
	- To prevent keep them filled with at least 3L fluid day & nimodipine IV/PO & insert central line to monitor central venous pressure – aiming for 8-10
	- Suspected with deteriorating GCS/ new neurological deficit
	- Treatment – Urgent CT brain to rule out a bleed as a cause of the deterioration then urgent angiogram to diagnose & treat vasospasm
	- Greatest risk of vasospasm is days 4-7 but significant risk for first 3 weeks after bleed, therefore will use preventative measures for at least 3 weeks
	Control of blood pressure
	- Main preventative measure is control of blood pressure – beta blockers
	Seizures
	- Prophylaxis with phenytoin
	Hyponatremia SIADH Cerebral salt wasting
Neurosurgical	**Coiling**
	- Endovascular technique done in angiography by interventional radiologists under GA
	- May be best if small necked aneurysm
	- Used in particularly sensitive areas e.g. basilar tip
	- Must be able to access the aneurysm (e.g. any stenosis or tortuous vessels)

Dr Nadeem Akhtar

	- Like dome:neck ratio to be 2:1 or greater
	Clipping
	- Craniotomy & careful dissection using microscope to reach aneurysm & clip usually at neck
	- May be performed after failed clipping
	- If aneurysm can't be reached by the endovascular root
	- levetiracetam
	- Ensure phenytoin levels are therapeutic
	- Treat as seizure from any cause & suspect re-bleed
	Treating hydrocephalus
Prognosis	- Re-bleeding
	- 80% mortality if re-bleed
	- Greatest risk is in the first 24 hours after the initial bleed
	- Aim to prevent by controlling BP to avoid dramatic changes & isolate the aneurysm from the circulation (coil or clip)
Pearls	- When blood enters the CSF (e.g. from SAH or during LP) the red cells are broken down & oxyhaemoglobin is released
	- It then takes 12 hours for the oxyhaemoglobin to be converted into bilirubin – conversion is via an enzyme found in the brain.
	- Bilirubin in the CSF, therefore, tells us that blood must have been in the subarachnoid space for at least 12 hours

- Blood which entered the CSF during the LP would not encounter the enzyme & could not produce bilirubin

- The CSF will look xanthochromic (yellowish discolouration) if bilirubin is present which they will look for with spectroscopy in the lab

Intracerebral Haemorrhage

Overview	▪ 10% ~ 15% of all cases of stroke
Risk factors	Age (> 55 years)
	▪ Men
	▪ Chronic hypertension
	▪ Amyloid Angioapthy
	▪ Heavy alcohol consumption
	▪ Cigarette smoking
	▪ Low serum cholesterol
	◆ Hypertension the most important!
	◆ Excessive alcohol use
	Common site
	▪ Cerebral lobe
	▪ Basal ganglia
	▪ Thalamus
	▪ Brain stem (pons predominantly)
	▪ Cerebellum
Clinical Features Signs	ICH
	◆ Hypertension (90%)
	◆ Altered mental status (50%)
	◆ Headache (40%)
	◆ Seizures (6-7%)
	Supratentorial ICH
	◆ Contralateral sensory-motor deficits involving putamen, caudate, thalamus, Aphasia, neglect, gaze deviation, hemianopia, subcorticle white matter or cortex

	Infratentorial ICH
	◆ Abnormal gaze, cranial nerve, contralateral motor deficits, brain stem.Ataxia, nystagnus, dysmetria, cerebellum
	Basal ganglla (50%)
	◆ Contralateral hemiparesis, sensory loss, conjugate gaze
	Lobar regions (20-50%)
	◆ Contralateral hemiparesis, sensory loss, aphasia, neglect, or confusion
	Thalamus (10-15%)
	◆ Contraiateral hemiparesis, sensory loss, gaze paresis
	Pons (5-12%)
	◆ Quadriparesis, facial weakness, decreased level consciousness
	◆ Cerebellum (1-5%)
	◆ Ataxia, miosis, gaze paresis
Investigations	▪ CT scan is the most effective tool in the ED with out contrast.
	▪ CT scan infarction or hemorrhage
	▪ Location and size of the hematoma
	▪ Presence of ventricular blood
	▪ Hydrocephalus
Management	
General Measures	▪ Intensive monitoring of neuroogic & cardiovascular status.
	▪ Instability is highest during the first 24 hrs GCS, hourly

Dr Nadeem Akhtar

Positional factors

- Avoid head and neck positions that compress jugular veins
- Avoid flat-supine position
- Elevate head of bed by approximately 30 degrees

Fluids

- Avoid hypo-osmolar fluids
- Replace urinary losses with normal saline
- Maintain euvolemia

Specific Measures	- Mannitol- Reduces cerebral edema by decreasing cerebral fluid volume, Rebound effect-use less than 5 days 20% solution 0.5-1.0 mg/Kg
	- Labetalol mean arterial pressure of 130 mm Hg20 mg IV, followed by 40 80 mg IV q10 min Titrate to BP or max 300 mgs
	- Pressors+ Avoid hypotension If systolic BP drops to less than 90 mmHg, consider judicious fluid boluses and / or start pressors
	- Prophylactic antiepileptic therapy
	- Consider prophylactic antiepileptic therapy in setting of ICH Lobar hemorrhage-35% seizure rate
	- Euvolemia Isotonic crystalloid solutions Elecrolyte abnormalities Correct deficits

Surgical indications	- Young patient with moderate or large lobar haemorrhage who are clinically deteriorating
	- Cerebellar Hameorrhage > 3 cm who are deteriorating over with brian stem compression and hydrocephalus form ventricular obstruction
	- Vascualar malformation if lesion is surgically accessible and patient has chance for good outcome
	- Small hemorrhages (10 cm3)
	- Minimal neurosurgical deficits
	- GCS< 4 (excluding cerebellar hemorrhage with brian stem compression)
	- Updated 25 12 14

Myasthenia Crisis

	Severe weakness of respiratory muscles, upper airway muscles (bulbar myasthenia) or both. often defined as FVC <1 L with the need for mechanical ventilation
Risk factors	- Poor control of generalized disease,
- High dose steroids
- Antibiotics, amikacin, gentamicin, streptomycin
- benzodiazepines,
- β-blockers)
- Botulinum toxin
- pregnancy
- systemic infections
- surgery.
Clinical Features	- severe dysarthria
- Dysphasia
- cough after swallowing
- Facial weakness,
- Weak palatal elevation, cough and gag reflex
- inability to raise the head due to neck muscle weakness
- paradoxical breathing.
- Test for muscle fatigue ability
- Look up and to the side for 30 sec
- If supine, look at feet for 1min
- Arms stretched forward for 1min
- 10 deep knee bends or squats
- Look for ptosis and ask about diplopia
- Absent cough and swallow reflex

MANAGEMENT

General Measures	Patients with myasthenia crisis can develop aponea suddenly and they must be closely observed
	<ul><li>Airway-open airway by suctioning secretions</li><li>Assess swallowing to prevent aspiration</li><li>Promote energy conservation measures</li><li>Administer high flow oxygen and measure oxygen saturation by pulse oximetry In patient without gag reflex, an oral airway may be placed.</li><li>Once the airway and breathing is stabilised the precipitating cause needs to be identified and removed.</li></ul>
	Discontinue offending Drug
	<ul><li>Anticholinesterases,</li><li>In patient if intubated or at risk for intubation, these medications are usually with held as they may increase secretions and potentially increase weakness in overdose,</li><li>Offending drug (e.g. antibiotic, B-blocker)</li></ul>
Specific Measures	Identify and treat infection
	IVIG
	<ul><li>Human intravenous immunoglobulin (0.4 mg/Kg/day * 5</li></ul>
	High dose corticosteroids
	<ul><li>Prednisolone 1 mg/Kg/day</li></ul>

Dr Nadeem Akhtar

	• Oral prednisolone at a dose of 1 mg/Kg for first 8-10 weeks tapering it by 5 mg every week, to potentially minimize the risk of steroid-induced myopathy.
	If respiration remains inadequate,
	• Assessment of severity with importance given to vital capacity and Bulbar function
	Contract anesthetist for endotracheal intubation
	• FVC monitoring if less then <15-20 mL/Kg;
	• Hypoventilation (pCO_2>45 or significantly increasing) or hypoxia pO_2<70 on room air)
	• Severe bulbar dysfunction or aspiration
Prognosis	▪ With advances in therapy and intensive care the prognosis have improved over the years
Additional Learning Points	Cholinergic Crisis
	• Increased weakness due to overdose of anti-acetylcholinesterace medications
	• If difficult from myasthenic crisis, consider tensilon test.
	• The SLUDGE syndrome (i.e., salivation, lacrimation, urinary incontinence, diarrhoea, GI upset and hyper motility, and emesis) also may indicate cholinergic crisis
	• Overmedication
	• Decreased BP
	• Abd cramps

	• N/V, Diarrhea
	• Blurred vision
	• Pallor
	• Facial muscle twitching
	• Constriction of pupils
	• Tensilon has no effect
	• Symptoms improve with administration of anticholinergics (Atropine)
Pearl	▪ Anticholinesterase medications (e.g. pyridostigmine, 30-60 mg * 5 per day orally; IV dose 1/30 oral dose)

Wernicke's Encephalopathy

Overview	Wernicke's encephalopathy is an acute neuropsychiatric syndrome resulting from thiamine deficiency, associated with significant morbidity and mortality. Continues to be an unrecognized and often misunderstood disease
Risk factors	• Thaimine deficiency (due to malnutrition or excessive alcohol intake)
Clinical Features	• The classic triad of mental status change, ophthalmoplegia andtruncal ataxia is present in only 16% of patients
Investigations	• No specific lab test
	• Complete blood levels
	• Serum glucose levels to exclude hypoglycemia or hyperglycemia
	• Pulseoxymetry &/ to exclude hypoxia/hypercarbia
	• Toxic drug screening
	• LP to exclude CNS infections
	Erythrocyte transketolase levels
	• Reliably detect thiamine deficiency butarenotnecessaryforthediagnosis of Wernicke encephalopathy.
	• The extent of thiamine deficiency is expressed in percentage. Normal values range from 0-15%; a value of 15-25% indicates thiamine deficiency, and greater than 25% indicates severe deficiency.7

Management	• Recommended that all comatose or hypothermic patients, as well as those with more classic presentations of Wernicke's encephalopathy, be given parenteral thiamine before administration of glucose.
	Pabrinex ♦ (Intravenous high potency) two ampoule pairs over 10 minutes which may be repeated eight hourly for two days, followed by one ampoule pair intravenously per day until the patient can tolerate oral thiamine. ♦ This should be continued at 100 mg bid for at least three months and until the patient stops drinking.

Dr Nadeem Akhtar

Transient Ischaemic Attack

Overview

TIA

The current definition of a TIA describes symptoms lasting less than 24 hours attributed to focal ischemia in a vascular distribution of the brain or retina.

- The majority of TIAs resolve within 60 minutes, and most resolve within 30 minutes.
- Less than 15% chance of complete resolution of symptoms if last >1 hour

Although implementation of the proposed tissue based TIA definition has not yet been adopted, this definition may more accurately reflect a true "Transient Ischemic Attack"

With 40-60% of TIA patients having evidence of ischemic injury on Diffusion Weighted Imaging (DWI),

Risk

Risk of stroke following a TIA is high:

- 10-20% within 90 days
- 50% of these within the first 48 hours

Clinical Features

Symptoms last less than 24 hours

- Most last less than one hour
- Less than 10 percent > 6 hours
- Amaurosis fugax up to five minutes

ABCD2 Score:

TIA

The current definition of a TIA describes symptoms lasting less than 24 hours attributed to focal ischemia i n a vascular distribution of the brain or retina. The majority of TI As resolve within 60 minutes, and most resolve within 30 minutes.

	Age 60 or older 1 point
	Blood Pressure >140/90 1 point
	Clinical
	-unilateral weakness 2 points
	-Speech impairment 1 point
	Duration
	-60 minutes or more 2 points
	-Less than 60 minutes 1 point
	Diabetes 1 point
	◆ Score 4 or greater – admit to hospital (moderate-high stroke risk).
	◆ Score predicted risk similarly among all ethnic backgrounds.
	◆ Best predictor of 2, 7, and 90 day stroke risk among validated scales.
General Workup	▪ Ecg ▪ FBC, Coags, and Electrolytes ▪ Chest Xray
CT scan	▪ Head CT without contrast ▪ Expedite if early presentation
MRI	▪ Distinguish new versus old ischemic areas. ▪ Distinguish new ischemic areas even with clinical TIA. ▪ Differentiate stroke etiology (small vessel vs. large vessel; embolic sources).

Carotid Imaging	Carotid Duplex Ultrasound
	• Sensitivity of 94 - 100% for > 50% stenosis
	• May over diagnose occlusion
	• Non-invasive
	Magnetic Resonance Angiography
	• Similar sensitivity to carotid ultrasound
	• Overestimates degree of stenosis
	• Gives information about vertebrobasilar system
	• Accuracy of 62% in detecting intracranial pathology
	• Cost and claustrophobia
	Cerebral Angiography
	• Gold standard for diagnosis
	• Invasive, with risk of stroke of up to 1%
	• For patients with positive ultrasound
	• For patients with occlusion on ultrasound
	• First test if intracranial pathology suspected
Management **General Measures**	▪ Neuro-obs; follow blood pressures.
	▪ Cardiac telemetry (paroxysmal a. fib).

Specific Measures	There are 2 proven therapies to prevent the occurrence of stroke following TIA
	♦ Antiplatelet
	♦ Anticoagulation therapy
	♦ Carotid Endarterectomy
Antiplatelet	Patients with TIA on ASA should have change in agent
	♦ Dipyridamole plus ASA
	♦ Clopidogrel
	Increase dose of ASA to 1300 mg/day
Anticoagulation therapy	If cardioembolic source
	♦ Long-term anticoagulation
	♦ INR acceptable range 2.0 – 3.0 (target 2.5)
Carotid Endarterectomy	♦ If TIA due to ≥ 50% stenosis in extracranial carotid artery consider CEA
	♦ Greatest benefit if surgery within 2 weeks

Dr Nadeem Akhtar

Stroke

Overview	Stroke (cerebrovascular accident, CVA): rapidly developing clinical signs of focal or global disturbance of cerebral function, with symptoms lasting 24 hours or longer, or leading to death, with no apparent cause other than a vascular origin
Risk Factors	<ul><li>Old age</li><li>Male Sex</li><li>Ethnicity</li><li>Prior stroke</li><li>Asymptomatic carotid stenosis</li><li>Sickle cell disease</li><li>Postmenopausal hormone therapy</li><li>Diet and nutrition</li><li>Physical inactivity</li><li>Obesity and body fat distribution</li><li>Hypertension</li><li>Diabetes</li><li>Dyslipidemia</li><li>Atrial fibrillation</li><li>Other cardiac conditions</li><li>Cigarette smoke</li></ul>
Classification	By vascular territory <ul><li>Ant. Circulation: carotid arteries</li><li>Post. Circulation: VB system</li></ul>

	By stroke etiology ♦ Atherothrombosis ♦ Embolus: ♦ Material: Red (fibrin rich) or White (platelet rich) ♦ Source: Cardiac? Aortic? Carotid Artery? ♦ Small artery disease ♦ Hypoperfusion: Hemodynamic ♦ Others: arterial dissection, arteritis, etc.
Clinical Features- by territory	Carotid territory ♦ Amaurosis fugax ♦ Dysphasia ♦ Hemiparesis ♦ Hemi-sensory loss Vertebrobasilar ♦ Hemianopia ♦ Quadraparesis ♦ Cranial N dysfunction ♦ Cerebellar syndrome ♦ Crossed deficit ♦ Loss of consciousness
Workup	♦ Brain imaging: CT, MRi ♦ Cardiac Imaging: TTE, TEE, heart monitoring ♦ Lipid, coagulation testing ♦ Vascular Imaging: ♦ Noninvasive

Dr Nadeem Akhtar

	<ul><li>MR angiography (MRA), intracranial, extracranial</li><li>CT angiography (CTA) Intracranial, extracranial</li><li>Ultrasound: Carotid, TCD</li><li>Invasive</li><li>Conventional cerebral angiograph</li></ul>
Neuro-imaging	CT Head<ul><li>Distinguishes reliably between haemorrhagic and ischemic stroke</li><li>Detects signs of ischemia as early as 2 h after stroke onset</li><li>Identifies haemorrhage immediately</li><li>Detects acute SAH in 95% of cases</li><li>Helps to identify other neurological diseases (e.g. neoplasms)</li></ul>Early signs of ischemic stroke:<ul><li>Hyper-dense vessel sign</li><li>Loss of insular ribbon</li><li>Obscuration of lenticular nucleus</li><li>Loss of gray-white matter distinction</li><li>Sulcal effacement</li><li>Areas of hypo-attenuatio</li></ul>

	MRI Brain
	
	- Diffusion-weighted imaging (DWI) :
	
	    ♦ Detects areas of restricted diffusion of water
	
	    ♦ Bright-up in acute ischemic stroke
	
	    ♦ Differentiation between new and old lesions
	
	- Perfusion-weighted imaging (PWI):
	
	    ♦ Detects abnormal tissue perfusion
	
	- Diffusion-perfusion mismatch:
	
	    ♦ Area of penumbra?

Indication of Thrombolysis	New ischemic stroke with well defined onset. time from first symptoms to t-PA <3 hours.
Contraindications	1. Evidence of intracranial hemorrhage on pretreatment CT 2. 18-Age 3. Symptoms rapidly improving or minor (not measurable by the NIH Stroke Scale). 4. Coma, or severe obtundation. 5. Pretreatment SBP > 185 or DBP > 110, despite simple measures 6. Abnomal blood glucose 7. Seizure at the onset of stroke 8. Known bleeding diathesis, including but not limited to: 9. (Platelet count < 100,000, current use of oral anticoagulants 10. PT > 15 sec, INR > 1.7,)

	11. Arterial puncture at a noncompressible site or LP within 7 days
	12. Major surgery or major trauma within 14 days
	13. GI or urinary tract bleeding within 21 days.
	14. Serious head trauma or previous stroke within 3 months.
	15. Past medical History of intracranial hemorrhage.
	16. Recent MI complicated by pericarditis.
	17. Pregnant or lactating females
Not a contraindication for thrombolysis:	▪ Current aspirin, NSAID, Ticipidine or Clopidogrel use.
	▪ History of PUD (not active).
	▪ Recent myocardial infarction.
	▪ Menstrution is not a contraindication
investigations	Head CT (noncontrast)/ MRI
	▪ To rule out haemorrhage. Significant hypodensity take about after 24-48h to develop. After 24 hours, if anticoagulant or antiplatelet therapy is to be given, a follow up CT scan or MRI should be free of hemorrhage. STAT Head CT for any worsening of neurologic condition
Treatment General measures	▪ Establish !V access at two sites, start 0.9% NS
	▪ Cardiac monitor, pulse oximeter, continuous vital signs.
	▪ STAT Labs: PTT, INR, CBC(without diff.),electrolytes, BNUN, creatinine, glucose, type & hold.

	MRI Brain • Gold standard DWI (diffusion-weighted imaging); bright signal with correlated dark signal. • ADC (apparent diffusion coefficient); appears with in 30 min, lasts for 1 week. • PWI (perfusion-weighted imaging) shows tissue at risk
Thrombolysis	**Alteplase** • Clinical S&S of definite acute stroke • Clear time of onset • Presentation within 4.5 hrs of acute onset • Haemorrhage excluded by CT scan • Age 18 and over • NIHSS less than 25 • Consent discussion
	Alteplase Dose • 0.9mg/kg/body weight, up to max of 90mg. • Diluted with sterile water to 1mg/ml • 10% of infusion as bolus • 90% as infusion using syringe pump over 1 hour

MONITORING	Risk of ICH is 6%, of which 3% are fatal - Vital signs and neuro checks: - Every 15 minutes for 2 hours after starting infusion. - Maintain sbp between 110 and 185mm Hg. - Insertion of a Nasogastric tube should be avoided, if possible, during the first 24 hours.
Avoid	- Foleys catheter - Central venous access and arterial punctures should be avoided. - Intramuscular injections should be avoided. - NBM except meds for 24 hours. - NO anticoagulants should be administered for 24 hours (including ASA, NSAIDs).
If hemorrhage is suspected	- Stop infusion of the thromboltic drug. - Stat-Head CT if ICH is suspected. - CT first to exclude hemorrhage. - 6 to 8 units of cryoprecipitate containing factor VIII. - Prepare for administrator of 6 to 8 units of platelets
Antiplatlet agent for first two weeks	- CT first to exclude hemorrhage - Early treatment with aspirin (150-300 mg/day) is recommended in acute ischemic stroke, should be started once ICH has been excluded on ct HEAD. this, reduces the risk of death and cardiovascular events Antiplatelet drugs:

	• Aspirin started within 48 hours reduces mortality and recurrent stroke.
If intolerant to to Aspirin/ If Aspirin is contraindicated	• Clopidogrel (75 mg/day) should be considered as an alternative to aspirin in suitable patients with contraindications to aspirin, or who are intolerant of aspirin, for prevention of cardiovascular events following ischaemic stroke.
If there is no NG access	• PR administration
Antiplatlet agent after first two weeks of stroke	• Clopidogrel 75 mg OD after first two weeks
Anticoagulation:	• Routine anticoagulation of stroke patient is not recommended.
	• In patients with atrial fibrillation, warfarin (target INR 2.5, range 2.0-3.0) should be used in preference to antiplatelet therapy to reduce the risk of a further ischaemic stroke because of its greater efficacy
	• No evidence that early anticoagulation reduces morbidity, mortality, or early recurrent stroke. Studies have shown that immediate
	• Treatment with heparin reduces DVT and PE but associated with increased risk of cerebral hemorrhage
	• In patients at increased risk of venous thromboembolism, additional prophylaxix with graduated elastic compression stockings should be considered in all immobile patients follo0wing acute stroke
Pearl	• Do not use low molecular weight Heparin-routinely

Dr Nadeem Akhtar

When to start anticoagulant therapy	Early institution or maintenance of anticoagulant therapy (with heparin or warfarin) in acute ischaemic stroke should be reserved

- For patients with a high risk of either venous thromboembolism (e.g. previous venous thromboembolism, thrpmbophilias)

- Or recurrent thromboembolic stroke

- e.g. rheumatic valve disease or mechanical heart valves, particularly in the presence of atrial fibrillation

- In situations where risks are judged to outweigh the increased risk of intracranial bleeding. Should also be considered as secondary prophylaxis after cardioembolic stroke from valvular heart disease or recent myocardial infarction

www.ingramcontent.com/pod-product-compliance
Lightning Source LLC
Chambersburg PA
CBHW030927180526
45163CB00002B/482